a history of science

a history of science

From agriculture to artificial intelligence

MARY CRUSE

This edition published in 2021 by Arcturus Publishing Limited
26/27 Bickels Yard, 151–153 Bermondsey Street,
London SE1 3HA

AD007734UK

Printed in the US

CONTENTS

PART FOUR:
The 19th Century

PART FIVE:
The 20th and 21st Centuries

PART SIX:
The Future of Science

PREFACE

Let's start with the simple stuff: what is the history of science?

Well, as it turns out, that's not actually such a simple question after all. In its most basic terms, the history of science is the study of how human beings have developed scientific theories, disciplines and knowledge over time. But, of course, what science is – the wider context and understanding of its purpose – has also shifted over time. Throughout the centuries, our conception of what constitutes 'science' has developed; from ancient natural philosophers and medieval alchemists to Renaissance scholars and Enlightenment reformers, we have travelled a long way to arrive at the modern conception of science and scientists that we recognize today.

Given that science is a fluid concept, it's not so surprising that the history of science is also a tricky topic to pin down. In his seminal 1962 text *On the Structure of Scientific Revolutions*, the American physicist Thomas Kuhn established the idea of the history of science as a series of paradigm shifts. Under Kuhn's model, scientific history is the process through which new theories gradually rise to prominence until they take over and replace the previous scientific paradigm; for example, quantum physics overtook Newtonian mechanics in the early 20th century. With this notion, Kuhn did away with the idea that the history of science represents a steady, cumulative march towards progress. In reality, the history of science is messy, plagued by flawed ideas and long periods of stagnation.

As we'll see, science itself is as complex and multifaceted as the problems it seeks to unpick. To make sense of it, we'll be looking at the process of knowledge-making chronologically and in the context of specific disciplines. But in order to talk in any broad sense about the history of science, it is necessary to omit large parts of the story, and there are many important and worthy topics and events that we're not able to cover within these pages.

The slipperiness of the subject matter, the vastness of the story – there are many considerations to bear in mind when exploring the history of science. But, perhaps most importantly of all, we must acknowledge that the history of science is not a static product; it is constantly shifting and evolving. This book would look very different if it were written in a few decades' time, and indeed, we'll be exploring some of the ways in which future historians of science might look back on the era we are currently experiencing.

Ultimately, the period in which we currently live is just another stop on the long timeline of scientific enquiry. As you read this book, you experience it amid a history that is still being created all around us, every day. To commandeer the words of the great American writer William Faulkner, 'The past is never dead. It's not even past.'

And so the story continues.

INTRODUCTION

'The most beautiful thing we can experience is the mysterious. It is the source of all true art and science.'

– ALBERT EINSTEIN 1879–1955, THEORETICAL PHYSICIST

Our story begins with a bang. It begins with the entirety of the universe clustered together; a hot, dense mass many times smaller than a pinhead. A fraction of a second passes, and the universe bursts free, expanding outwards into immensity. It continues to grow. Stars develop and die; entire galaxies are formed; solar systems spring up, and from the grand chaos that is the cosmos, our own little corner of the universe is born.

And that is where our story truly begins. On a little blue planet nestled in the Milky Way galaxy, a complex set of chemical phenomena align in just the right way so as to create something miraculous: life. Fast forward a couple of billion years, and one group of creatures splits into two. A succession of animals follow, each one looking more like the face you see in the mirror – *homo habilis*, the handy man; *homo erectus*, the upright man – until finally, around 300,000 years ago, *homo sapiens*, the wise man, is born.

We *homo sapiens* live up to our name. We are defined by our capacity for wonder, and for the short time that we have been on the planet we have stared up at the vast, infinite expanse of darkness above and have been filled with questions.

To be human is to wonder. This is our legacy as members of the human race. The impulse to ask questions is hardwired into our DNA, and for three hundred millennia our people have been searching for answers. Carl Sagan, the great American astronomer and science communicator, famously said that we are all made of star stuff, because the atoms that make up the human brain also make up the stars. And so, in a sense, the human brain is the universe attempting to understand itself, and humanity's way of understanding the universe is science.

The history of science is the history of human endeavour; a testament to our will to discover and our determination to uncover truth and meaning amidst the multitude of unknowns. Of course, that history is by no means neat. There are rapid accelerations and sudden stops, there are periods of silence, wrong turns and disagreements. But over the course of our journey, the wise man has slowly chipped away at the vast mysteries that surround us.

In the early days of our story, we did this by painting what we saw around us on to cave walls; these days, we do it by splicing genomes, sending rovers to Mars and smashing particles together in giant colliders. We still don't have all the answers; there is still so much left to discover. But if there is a common thread that runs through the entire history of science, it is this: all that we know about the world is the product of humanity's capacity for wonder.

You are part of this story, because you share in the spirit of curiosity that drives scientific progress. Our innate sense of wonder is what makes us human, and it is what makes us – all of us – scientists. And so, before you delve into the history of science, know this: the story is not over yet. It continues with all of us; it continues with you.

PART ONE

ANCIENT HISTORY

3000 BC–5TH CENTURY

WHEN WE THINK ABOUT SCIENCE, we tend to turn to familiar tropes: futuristic machines, complex experimental set-ups, people in lab coats. But science has existed for far longer than any of that; in fact, you could say that human beings have been doing science in some form or another for thousands of years.

If we understand science to mean the pursuit of evidence-based knowledge, then it's clear that humanity and science go back well before historical records began. So let's take a look at the early days of civilization and the years leading up to it. In place of what we call science, another belief system prevailed. Early civilizations perceived a universe in which events were made to happen by deities, or in which spirits and energies governed the natural world. While they might not have understood the scientific method, they probably still questioned the world around them and created theories for how things worked. So what were some of the earliest examples of 'science' as we understand it today, and what part did the scientific method play in the important developments and transitions that underpinned human progress?

We've established that humanity has been asking questions for as long as our species has existed. Now, let's look at how we began using science to find answers.

The era of ancient history begins with the advent of writing in 3200 BC, and ends around 476 BC with the fall of the western Roman Empire. We tend to associate science in this period with Ancient Greece, but it was actually a global endeavour. From India to Egypt to Mesoamerica, the ancient period saw different groups of people begin making scientific progress in different areas and in different ways. One quality that united the early scientists – whether they were philosophers in Ancient Greece or medics in Aryuvedic India – was their desire to seek order in nature, and it was around this time that the first scientists began looking for consistent laws and structures that could explain the phenomena they observed around them. This wasn't science as we imagine it today. Protoscientific ideas were frequently – though not always – intermingled with religion and mysticism. But this type of thinking did represent an important shift away from the use of observation and logic solely to achieve practical goals, towards the pursuit of knowledge for knowledge's sake. People were no longer content to merely observe and exploit the natural environment and physical phenomena; they wanted to *understand* them. And as they began to acquire a deeper understanding of the world, they became better able to manipulate the environment around them, contributing to the rise of larger and more powerful civilizations. Science is fundamental to the story of human civilizations, and it always has been, right from the very start.

CHAPTER 1
MATHEMATICS

'Wherever there is number, there is beauty.'

– PROCLUS DIADOCHUS, GREEK PHILOSOPHER

Mathematics is the science of numbers, quantities, measurement and shapes. But it's not a science in the way that biology or physics are sciences. Rather than using the scientific method, which is based on observation, theories and evidence, maths is based purely on logic; so its findings are abstract, not tied to the physical world. But while maths isn't quite like the other sciences, it is fundamental to scientific enquiry. From microbiologists to astrophysicists, all sorts of different scientists use maths as a tool to explore the world. Mathematics has supported and strengthened civilizations around the globe. From its origins as a tool in Mesopotamia and Egypt to its maturation into a scientific discipline in Ancient Greece, the history of maths is long and complex, but its impact on our daily lives is completely clear. When we think of maths, we might initially envisage fractions and long division, but there's so much more to it than that. Mathematics has helped researchers to better understand the world around us, and has played a part in a whole host of innovations – from architecture, to aeroplanes, to androids. And none of that would have been possible without the ingenuity of ancient peoples who, thousands of years ago, imagined mathematics into existence.

ANCIENT MATHEMATICS

Although it has since become a discipline of study, mathematics was originally born out of necessity. The first written evidence comes from the Sumerian peoples of Mesopotamia, in what is now mostly modern-day Iraq. This civilization – which lasted from around 4500 to 1900 BC – used maths as a tool to support their emerging agricultural society. Around the third millennium BC, the Sumerians began practising geometry and multiplication, carving their calculations into wet clay. From taxing harvests to measuring plots of land, mathematics helped the Sumerians to quantify and organize material.

Did You Know?
The Sumerian, Mayan and Indian civilizations all independently invented the concept of zero.

A later Mesopotamian civilization, the Babylonians, further developed mathematics. They were the first to design a positional number system, in which the position of the number denotes its meaning. We still use a positional number system today – the number 5 written alone denotes only 5, but when it precedes another number, it comes to signify 50; when it precedes another two numbers, it comes to mean 500. The Babylonians used their sophisticated mathematical knowledge to help chart the stars, allowing them to predict lunar eclipses and planetary cycles, and to create their own 12-month calendar. Having a calendar enabled better control over agricultural seasons and supported the growth of religious occasions and feast days.

c. 3000 BC	*Sumerians begin practising geometry and multiplication*
c. 6th CENTURY BC	*Pythagoras establishes society of mathematicians*
c. 3rd CENTURY BC	*Euclid organises and shares fundamental rules of mathematics*
c. 3rd–4th CENTURY BC	*First recorded use of zero*
c. 3rd–2nd CENTURY BC	*Archimedes accurately calculates value of pi*
c. 6th CENTURY	*Hindu-Arabic numeral system is developed in India*

9th CENTURY	*Al-Khwarizmi publishes treatise on elementary algebra*
1637	*René Descartes develops Cartesian geometry*
1837	*Charles Babbage proposes Analytical Engine*
1854	*George Boole fully develops true-false system of algebra, used in computing*

The Ancient Egyptians' grasp of mathematics was less advanced than the Babylonians, but still left a mark on its history. Mathematical activity in this region was largely concentrated around the activity of scribes: young men who were among the few Egyptians trained to read and write. The scribes were the civil servants of their time, practising accountancy, note and letter writing, and many other administrative activities that required mathematical knowledge. Scribes used hieroglyphs to represent numbers, employing a decimal scheme that revolved around the number 10. However, the Egyptians didn't have a positional system, which meant that they had to count each number individually. So 600 would be represented by drawing the sign for 100 six times. This method was laborious and inefficient, but it did provide a uniform system for calculating large numbers. It's likely that the Egyptian system of mathematics influenced Ancient Greeks – like Thales and Plato – who visited Egypt and brought ideas back home with them. This impact is important, because the Greeks would go on to embrace the science of mathematics like never before.

Did You Know?

An abacus is an ancient calculating device that was probably invented by the Babylonians. Modern abacuses mostly work by sliding counters across rods on a frame, but the humble abacus has seen many different iterations in various cultures over the centuries.

While the Greeks were by no means the first thinkers to explore mathematical concepts, it was in Ancient Greece that mathematics

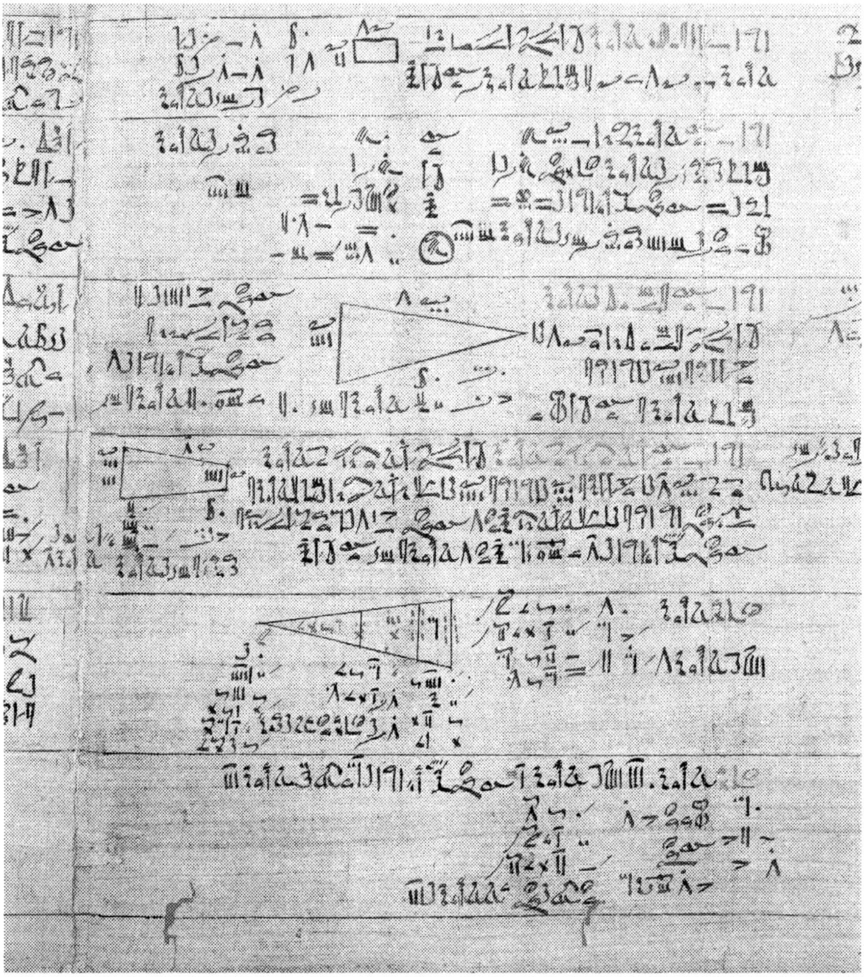

The Rhind papyrus is one of the best-known examples of Ancient Egyptian mathematics.

became a discipline in its own right. The Greeks were also the inventors of the word 'mathematics', from the ancient Greek word *máthēma*, meaning 'that which is learnt'.

Thinkers were interested in mathematical concepts rather than what maths could be used for. They devised the concept of 'proofs': logical formulae which demonstrate that a mathematical rule – such as $a^2 + b^2 = c^2$ – is always true. By proving the existence of solid mathematical laws, the Ancient Greeks established maths as an independent field of study and a way of understanding the world.

Pythagoras and his followers were among the first philosophers to fully embrace mathematics in the 6th century BC. The motto of the Pythagoreans is said to have been 'all is number', and they believed that numbers were not only important but also sacred. They ascribed religious meaning to certain numbers, and held unusual beliefs – for instance, it's claimed that Pythagoreans would never gather in groups of more than 10, because they considered 10 the perfect number. Despite these quirks, the Pythagoreans are credited with some genuine mathematical contributions, including Pythagoras' Theorem: a formula that can be used to work out the length of one side of a triangle based on the length of its other two sides.

A bust of the Greek philosopher Pythagoras (c. 569–475), inventor of the formula $a^2 + b^2 = c^2$.

By the dawn of the 3rd century BC, mathematics had become a central field of study in Ancient Greece. It was around this time that the great mathematician Euclid published his seminal work *The Elements*, a series of books which laid out the fundamental mathematical rules that were understood at the time. While these rules were already known, Euclid was the first to organise and share them, and his treatise went on to become one of the most influential mathematical works in history.

However, the title of 'greatest mathematician of all time' is generally bestowed upon one of Euclid's successors, Archimedes. Born around 287 BC, Archimedes is famed for his extensive knowledge of geometry and mechanics. Over the course of his life, Archimedes proved theoretical methods for calculating the area and volume of various shapes, accurately calculated the value of pi, and established the laws of buoyancy.

Pi is the ratio of a circle's circumference to its diameter. Whatever the size of the circle, pi remains the same: 3.14159265… and so on. It wasn't called 'pi' until the 18th century, when Welsh mathematician William Jones named it after the letter P in the Greek alphabet. He chose P because it's the first letter of the word 'perimeter'.

Although thousands of years have passed, the Ancient Greeks' contributions to mathematics are still considered remarkable today. The very existence of mathematics as an academic discipline is the result of this civilization's fascination with numbers.

Who's Who – Hypatia

Hypatia was an astronomer, mathematician and philosopher born in Alexandria somewhere around the second half of the 4th century. One of the most accomplished scholars of her time, Hypatia was a revered figure in Ancient Greece. She published commentaries on geometry and arithmetic, and played an integral role in preserving the great works of the scientists who came before her, including the polymath Ptolemy and the mathematician Euclid, during a time of political and religious unrest in Alexandria. Because of her reputation for wisdom, Hypatia grew to be a trusted council to the Roman governor of Alexandria. Unfortunately, she became caught up in a power struggle between the governor and the bishop of Alexandria, and was ultimately murdered by a Christian mob. The brutal manner of her death prompted mass outrage, and led to Hypatia becoming venerated as a martyr for philosophy. She is still remembered today as one of the great thinkers of her time, and a symbol of science and enlightenment in the face of oppression and zealotry.

THE ROAD TO MODERN MATHEMATICS

The mathematical progress made during ancient times represents just a fraction of the scholarship in this vast field. In the years after Ancient Greece was conquered by the Romans, mathematic progress slowed. The Romans were fantastic engineers, and used mathematics to support the development of extraordinary machines and infrastructure, but it was in the Islamic world that the discipline flourished.

Islamic mathematical progress in the Middle Ages was influenced by the work of Indian scholars. The decimal system was used in India from the early 7th century, and it was here that the Hindu-Arabic numeral system originated. Islamic mathematicians were also heavily influenced by Greek mathematical scholarship, having translated and preserved the classical mathematical texts, but they did also generate significant contributions of their own.

One of the most prolific Islamic mathematicians was the Persian scholar Muhammad Ibn Musa al-Khwarizmi. Born in AD 780, Al-Khwarizmi lived in Baghdad and worked at the renowned House of Wisdom (see page 54), where he was a revered thinker. Al-Khwarizmi wrote the book on algebra, literally. His treatise on elementary algebra gave the discipline its name, and his own name later became the basis for the word 'algorithm'. In addition to broadening the world's vocabulary, Al-Khwarizmi's work also introduced the Hindu-Arabic numeral system and associated arithmetic to the West. In fact, it could be said that Al-Khwarizmi is one of the most influential figures in the history of mathematics, and his ability to apply his mathematical understanding to other fields such as geography and astronomy helped to advance scientific knowledge both in the Islamic world and beyond. In the post-classical era, Islamic mathematicians became the foremost thinkers in the field. But by the 15th century, the centre of mathematical scholarship was beginning to shift back towards Europe, where it would go on to flourish in the coming centuries.

The Hindu-Arabic Numeral System

Between the 6th and 7th centuries, Indian mathematicians developed the Hindu-Arabic numeral system, the number system that most of the world still uses today. This system uses 10 different numerals, counts in 10s and uses the position of the numeral to denote meaning. The Indian mathematicians' design was unlike any system of counting that had gone before, and transformed mathematics.

But it was through the Islamic world that the Hindu-Arabic numeral system spread to the West. The treatises of Middle Eastern mathematicians Al-Khwarizmi and Abu Yusuf al-Kindi were read by European scholars from around the 12th century, spreading the revolutionary number system further around the world. The versatility of the Hindu-Arabic numeral system made it a powerful tool for science and a cornerstone of scientific enquiry from the post-classical era onwards. While the number system we now know may feel completely natural, it's no accident; the numbers we know today are the product of centuries of hard work and consideration by ancient scholars.

Mathematics became more widely studied in Europe during the Renaissance, spreading out from universities and into the offices of merchants and traders, and even to artists' studios, where it was used to

*The Persian scholar Al-Khwarizmi (*c. *780–850), whose name became the basis for the word 'algorithm'.*

add new layers of perspective and symmetry in paintings. The field was soon swept up in the scientific revolution of the 16th and 17th centuries, as modern ideas about science and scientific thinking developed, and science grew into an established discipline. This period saw a surge in mathematical outputs and achievements in Europe. Mathematical ideas helped thinkers such as Galileo Galilei and Johannes Kepler to advance the science of astronomy, and allowed Isaac Newton to describe the laws of physics.

The French polymath René Descartes is recognized as one of the foremost mathematical thinkers of the 17th century. His work helped to connect the fields of algebra and geometry through the development of a new type of geometry known as 'Cartesian geometry'. Descartes is also thought to have established the use of 'x' in algebraic equations to represent an unknown quantity. During the 18th century, the Swiss mathematician and physicist Leonhard Euler became one of the most productive mathematical minds of his generation, publishing works that spanned the many branches of mathematics, including algebra, geometry, calculus and trigonometry. With his devotion to abstract mathematical concepts, Euler helped to develop the burgeoning field of pure mathematics.

Pure mathematics is the abstract science of mathematical concepts, as opposed to applied mathematics, which is focused on the practical uses of mathematics.

By the 19th century, we see the first signs of the fundamental role that mathematics will play in the modern world. In the 1830s, the English polymath Charles Babbage invented his 'Analytical Engine', an early precursor to the modern computer, capable of performing calculations and storing numbers in a memory unit. In the mid-19th century, the English mathematician, George Boole, devised a system of algebra that deals exclusively in true or false statements and can be applied to logical problems; 'Boolean algebra' would later become fundamental to modern computer science. At the same time, a number of mathematical societies began springing up around the world, from England to Italy to the United

States. By the end of the century, mathematics was firmly established as its own discipline, entirely distinct from the physical sciences.

Did You Know?

At the dawn of the 20th century, the German mathematician David Hilbert set out a series of 23 unsolved mathematical problems. The Hilbert Problems were considered the greatest mathematical quandaries of the era, and established a challenge for future generations of mathematicians. Of these problems, ten have since been solved and a further seven partially solved. Four are considered too ambiguous to fully solve. The final two remain open.

The Analytical Engine, designed by Charles Babbage.

Mathematics has continued to grow in depth and scope over the course of the 20th and 21st centuries, ultimately becoming a cornerstone of cutting-edge research. Mathematical modelling and techniques are used in diverse fields, from theoretical physics to drug design, and maths is a core component of data collection and analysis in modern experiments. Away from the sciences, mathematics is more central to our everyday existence than ever before. The rise of computer technology, together with the steady spread of that technology into more and more areas of our lives, has made mathematics one of the most important disciplines of modern times.

In the centuries since it was first developed, mathematics has allowed for the growth of innovative infrastructure and machines, supported developments in science and research, and provided thinkers with a deeper understanding of numbers and the way in which they shape the world. This was true thousands of years ago when the first mathematical ideas were being developed, and it's just as true now. From the technology and engineering that we depend upon, to medical advances and space exploration, much of modern innovation and invention depends on mathematics and the work of mathematicians throughout the ages.

CHAPTER 2
MEDICINE

'Medicine is of all the Arts the most noble.'

– HIPPOCRATES, GREEK PHYSICIAN

Medicine is the science and practice of disease diagnosis, treatment and prevention. Contemporary medicine is now more advanced than ever before, and the medical field is interwoven with a number of different scientific fields. In later chapters, we'll be exploring modern branches of medical research, including genetics, pharmacology and microbiology, but in this chapter we're taking a broad overview of the medical field as it developed in ancient cultures and beyond.

From Greece to China, India to Rome, the roots of medicine go deep into ancient civilizations around the world, branching out in all sorts of different directions. But this isn't medicine as we know it – ancient medical treatments looked very different to the care we receive today. Folklore and magic were influential in early civilizations and were intimately woven into the treatment of maladies. What's more, the lack of scientific tools and methods meant that much of what we knew about the body and its response to disease was based on observation combined with trial and error. Nonetheless, ancient peoples did have strategies for taking care of their health, and some of the treatments practised in ancient times did have therapeutic value. Ultimately, the ideas that were developed in the early days of medicine have helped to shape medical practice as it exists now. From the snake and staff that symbolize healthcare in much of Western culture to our enduring reverence for the Hippocratic oath, the influence of medicine's origins can still be seen and felt today. This is the story of how medicine developed in different ways within several early civilizations; it is also the story of how human beings came to better understand their bodies and the conditions necessary for health and survival.

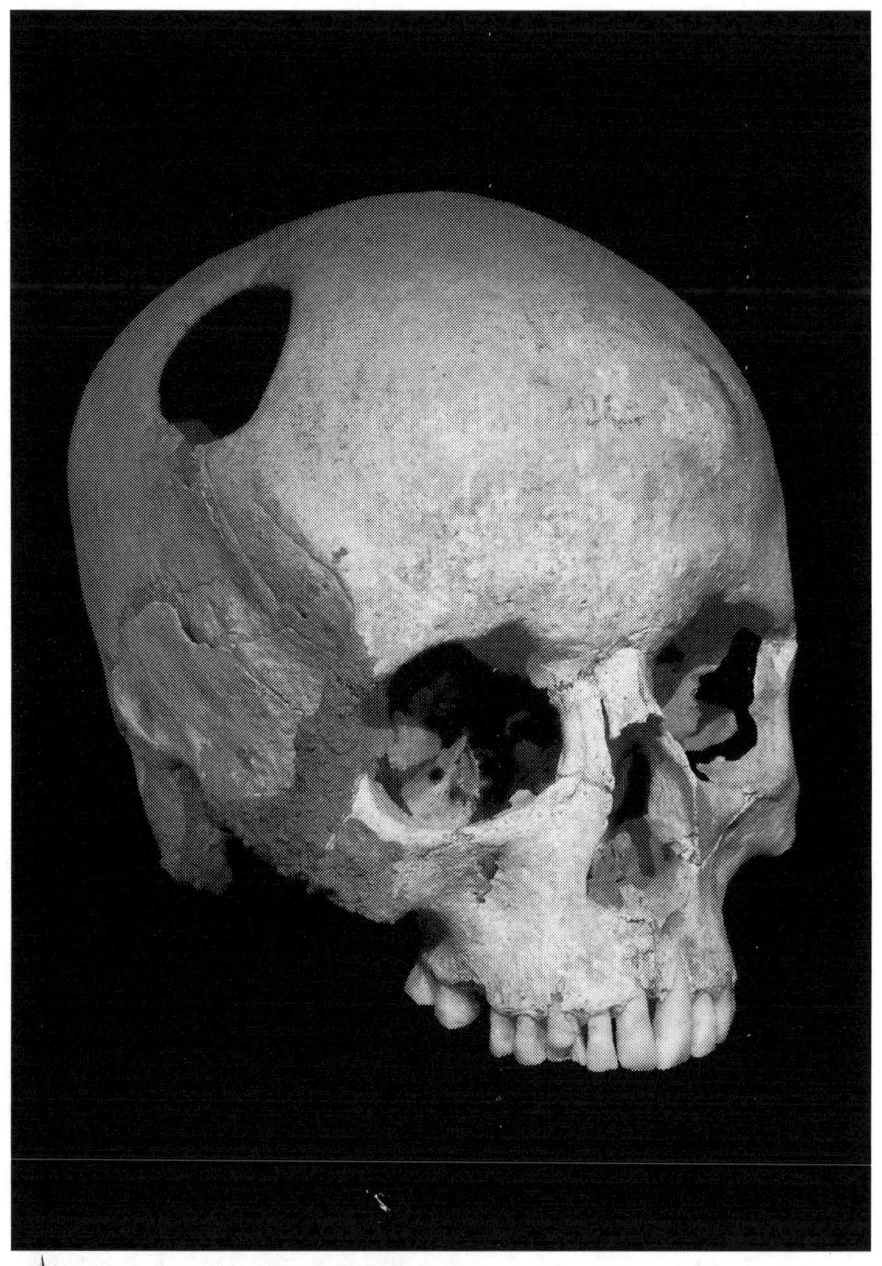

An ancient skull with evidence of trepanation.

c. 2000 BC	*Origin of the Vedas: sacred Indian texts containing medical information*
c. 1700 BC	*Babylonian tablet on medical case law*
c. 1600 BC	*Ancient Egyptians author the Edwin Smith Papyrus*
c. 5th CENTURY BC	*Hippocrates promotes rational medicine, dismissing divine influence*
c. 300 BC	*Bian Que uses general anaesthetic to sedate surgical patients*
19 BC	*First public baths, called thermae, introduced in Ancient Rome*
1025	*Ibn Sina completes his five-volume medical encyclopaedia*
16th CENTURY	*Practice of inoculation is widely used in China*
1849	*John Snow founds discipline of epidemiology*
1940	*Ignaz Semmelweis promotes importance of medical hygiene*

ANCIENT MEDICINE

Archaeological evidence suggests that human beings were practising medicine as far back as the Stone Age. Human remains discovered from this period suggest that broken bones were set and wounds stitched and healed. The practice of trepanning is also evidenced in archaeological findings; this was an often-fatal practice in which a hole was made in the skull.

For many years, medical wisdom for treating disease was passed down orally from one generation to another, but the advent of writing in Mesopotamia in 3500 BC allowed medics to record their knowledge directly for the first time. A Babylonian slab dating back to the 18th century BC contains case laws describing surgeons' pay and the consequences of malpractice, which included amputation of a surgeon's hands should they kill a patient.

Ancient Egypt was one of the first places to practice what we might recognize as medical care. The Ancient Egyptians were famed for their advanced medical knowledge. They were pioneers in surgery and dentistry, and even promoted the importance of good diet and nutrition. While the Ancient Egyptians were advanced in many aspects of their medical thinking, magic and mysticism remained fundamental to many of their therapeutic practices and beliefs. Doctors also acted as magicians, reciting spells and incantations designed to call upon the gods for assistance in healing their patient. Despite this, the Ancient Egyptians were renowned as a healthy and medically advanced civilization, and the practices they developed would go on to influence the medical landscapes in Greece and Rome.

The Edwin Smith Papyrus

One of the reasons we know about Ancient Egyptian medicine is because of a series of medical texts written on papyrus and discovered in the 19th century. These documents contain medical observations and advice on everything from basic anatomy to gynaecology. One such text, the Edwin Smith Papyrus, is an ancient manual for the surgical treatment of trauma wounds and the oldest known document of its kind in the world. Named for the dealer who purchased it in 1862, the papyrus dates to around 1600 BC, and is thought to be a copy of an even older document. Unlike other texts of its time, the Edwin Smith Papyrus is grounded in rational observation rather than magic. It describes 48 different cases of traumatic injury – from open head wounds to cut lips – recommending the appropriate examination procedure, and providing diagnoses, prognoses and potential treatments.

The papyrus demonstrates an extraordinary level of knowledge for the time at which it was written. Here, the word 'brain' appears for the first time in any language, and the writer shows an understanding of the impact of various brain injuries on movement. The work also discusses the relationship between the pulse and the

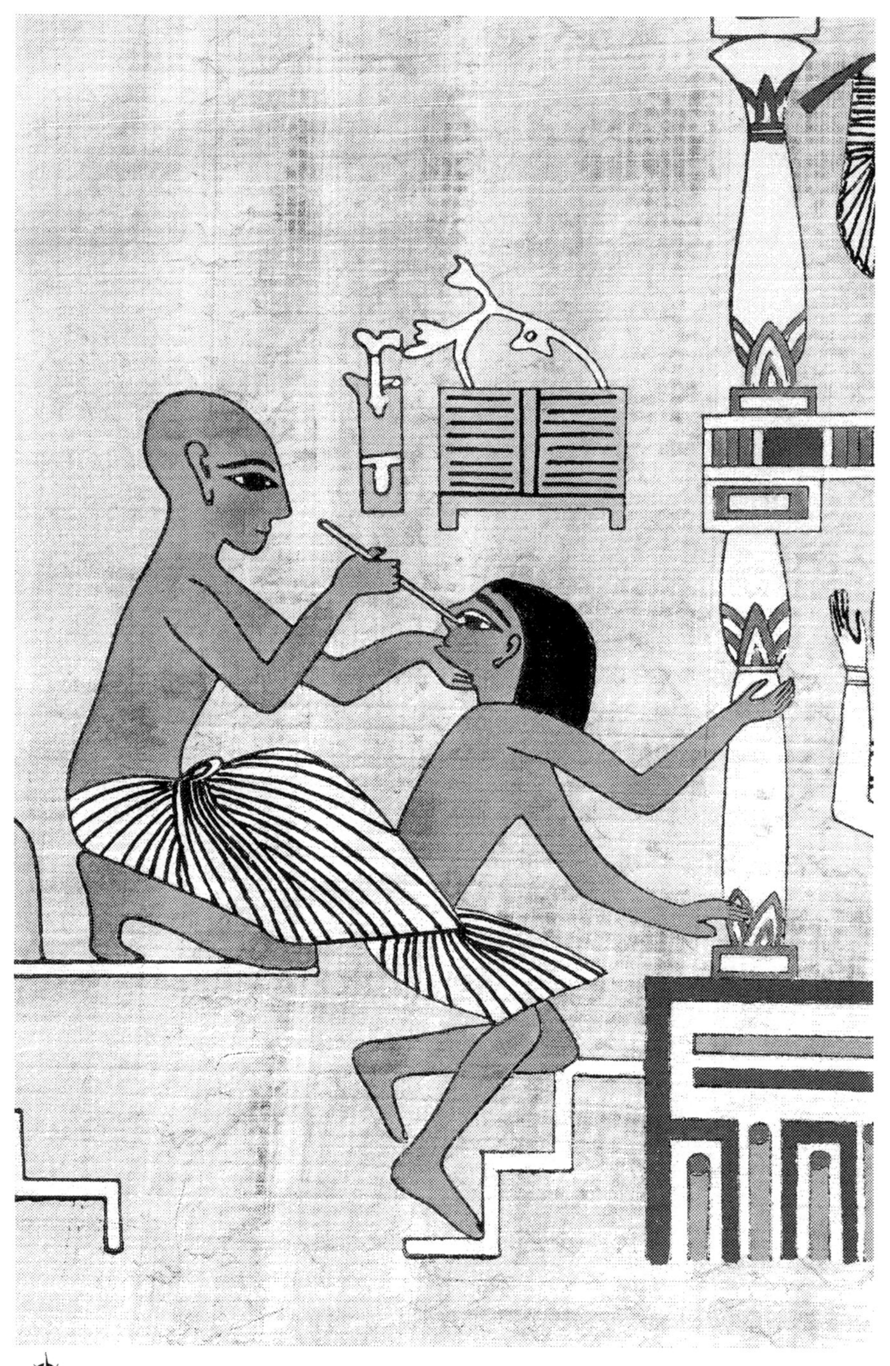

An Ancient Egyptian papyrus depicting ophthalmic medical treatment.

heart, along with the workings of some internal organs. At over 15 feet long, the Edwin Smith Papyrus provides a comprehensive insight into early surgical procedures in Ancient Egypt and demonstrates the depth of knowledge this civilization possessed.

Asian civilizations also played an important role in shaping the course of medical care. Early forms of medicine in India date back thousands of years, with some principles set out in the Vedas: sacred writings that originate in the 2nd millennium BC. These ancient texts tell us a lot about the practice of medicine in early India, and contain information on anatomy, medical knowledge and treatments. Ancient Indian medics were particularly skilled in the art of surgery. From amputations to caesarean sections, surgical procedures in India were notably advanced, while some of the earliest records of reconstructive surgery date back to this period. At this time, convicts sometimes had their nose forcibly amputated as a punishment for some crimes, including theft and adultery. At the same time, Indian surgeons learned to reconstruct the nose using skin grafts from other areas of the face, thus introducing the practice of rhinoplasty to the world. Thought to have been published prior to the 2nd century BC, the *Charaka Samhita* – a Sanskrit text on Ayurveda, traditional Indian medicine – lists hundreds of muscles, bones, joints and vessels in the human body and demonstrates the depths of Ancient Indian medical knowledge. The systematic and evidence-based approach to medicine in Ancient India meant that, at this time, Indian medicine and surgery were possibly more advanced than any other contemporary ancient civilizations.

Sanskrit is one of the world's oldest languages, and is considered the root of many modern languages.

Traditional Chinese Medicine developed independently from the Western tradition, but has parallels. Built on the idea that human beings are connected to larger cosmic principles of order and balance, practitioners could help resolve disharmony in human health through treatments such as herbal medicine and acupuncture. As early as 300 BC, the revered Chinese surgeon Bian Que was using general anaesthetic to sedate patients for surgical procedures, and by the 2nd century BC, the great physician Zhang Zhongjing had published his seminal work recommending treatments for typhoid and other fevers.

Did You Know?

The Ancient Chinese invented the practice of inoculation – a predecessor to modern vaccines. Inoculation works by introducing a live pathogen to a patient in a controlled way, so as to stimulate an immune response.

The Ancient Greek tradition is widely considered the dawn of Western medicine, although the Greeks were influenced by centuries of medical enquiry that took place all over the world in the years prior. In the early days of Greek civilization, illness was thought to result from the actions of deities, and so the sick would appeal to the gods for healing. However, the early philosophers, with their emphasis on observation and logical thought, helped to gradually shift the focus of medicine away from magic and towards science.

The most notable Greek physician, Hippocrates, is credited with much of this shift towards rational medicine. Writing on epilepsy, which was then thought of as a 'divine disease' sent from the gods, Hippocrates said, 'It is not any more sacred than other diseases, but has a natural cause, and its supposed divine origin is due to human inexperience. Every disease has its own nature, and arises from external causes.' This was a revolution in thought. If disease had external causes, then it could be treated with human action rather than by divine influence.

Although they were advanced in their medical thinking, the Ancient Greeks didn't get everything right. They believed that the cosmos was composed of four elements – earth, air, fire and water – and the body composed of four humours: phlegm, blood, black bile and yellow bile.

It was thought that the four humours must be kept balanced in order to maintain good health. Unbalanced humours, and the resulting disease, could be treated using a variety of methods, depending on which of the humours were affected – for instance, the blood balance might be restored by bloodletting, while the bile balance could be dealt with by purging. This theory lasted for almost 2,000 years, and heavily influenced early Islamic and medieval European medicine.

Who's Who – Hippocrates

Hippocrates is one of the most revered figures of Ancient Greece, and with good reason. More than virtually any other thinker that came before him, Hippocrates emphasised the importance of reason and observation in medicine. He was adamant that it was the environment, not the gods, that caused disease, and that it was within human control to address maladies.

While we know that he was born on the island of Kos somewhere around 460 BC, Hippocrates remains a figure shrouded in mystery. Very little contemporary writing on him exists, and much of the literature is embellished. Some years after his death during the Hellenistic Period, Hippocrates became a figurehead for Greek medicine, and over time scholars began to credit him for work done by other scholars. However, while Hippocrates wasn't responsible for everything commonly ascribed to him, he did make a huge contribution to Ancient Greek culture and to medical thought as a whole.

Hippocrates believed the prevailing theory at the time: that the human body was made up of four humours. While this assumption was clearly flawed, the treatments Hippocrates recommended in response to 'imbalances' – including diet, rest, exercise and herbal medicine – were useful and based on observation. Hippocrates and his followers were also the first to identify and describe a large number of different diseases, and to categorize illnesses as acute, chronic, endemic and epidemic.

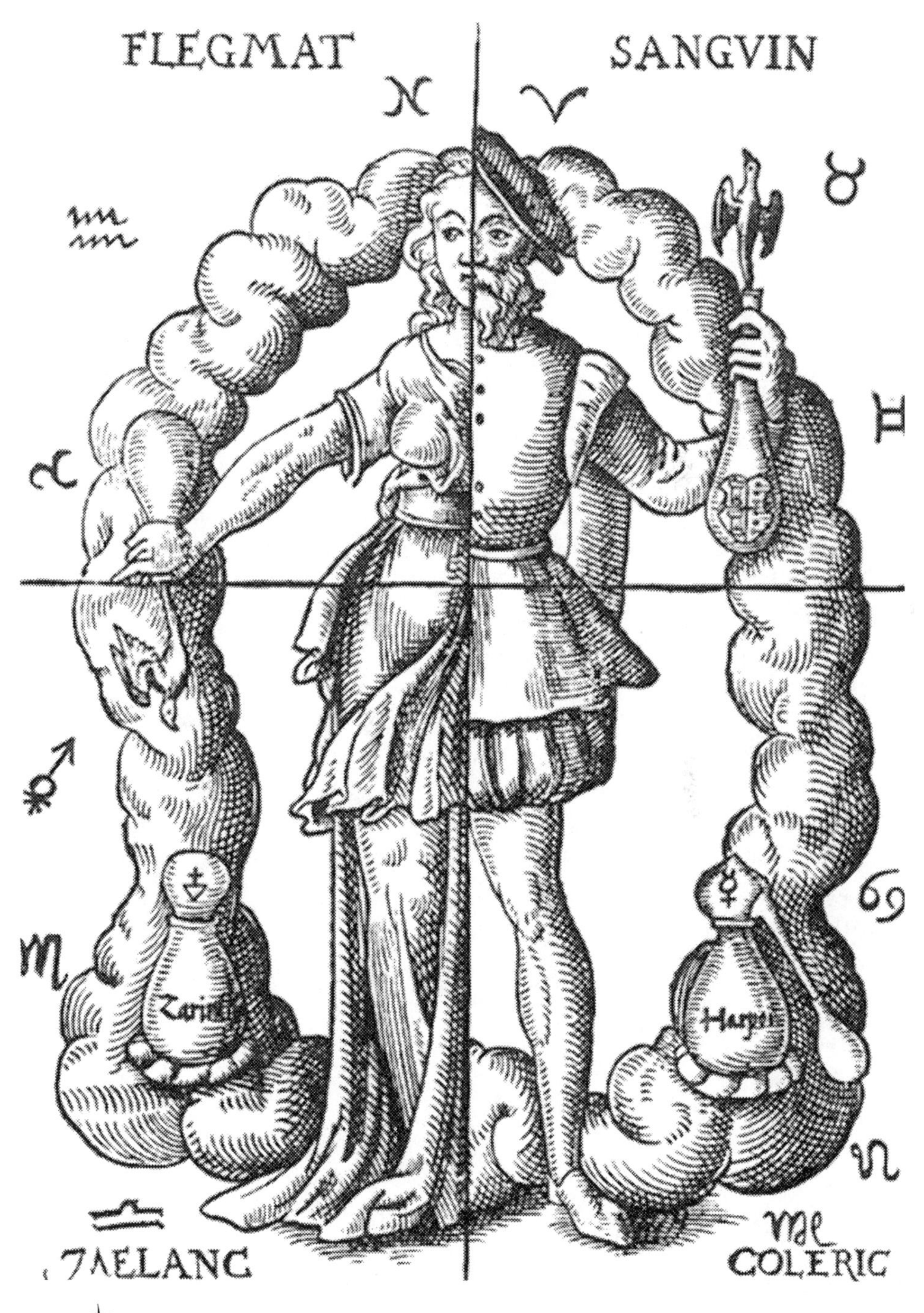

An artistic interpretation of the four humours: phlegm, blood, black bile and yellow bile.

Hippocrates is perhaps best remembered for his professional approach and commitment to treating patients according to a strict code of ethics. While he probably wasn't the originator of the Hippocratic Oath (like many of the contributions ascribed to him, this was actually created by other unknown scholars), Hippocrates' medical practice prioritized the welfare of the patient, taking a humane approach to treatments. For this, and for his rational, systematic ideas about medicine, Hippocrates will be forever remembered as one of the greatest physicians of all time.

In 146 BC, the Ancient Greek empire was finally conquered by the Romans, but this was not the end for their medicine. The Ancient Romans adopted many Greek teachings and practices; that said, they were also innovators in their own right. The Ancient Roman approach to public health was revolutionary. Good hygiene was considered extremely important and public baths were popular. Ancient Rome was also home to some of the earliest hospitals, built initially to treat soldiers and veterans of war. The Romans' skill in engineering also impacted on their health, with aqueducts supplying clean water and advanced sewage systems improving sanitation. The Romans' commitment to public health was well ahead of its time, and it would be many centuries before this sort of infrastructure was widely adopted elsewhere. What's more, in the centuries to come, medical progress would stagnate in the medieval West, as knowledge was lost and medicine remained deeply entwined with religion.

THE ROAD TO MODERN MEDICINE

Although heavily influenced by the Greco-Roman tradition, the approach to medical research in medieval Europe looked different from what went before. The medical contributions of ancient scholars such as Hippocrates and Galen were generally accepted without question, which stifled progress in medical thought for some time. Meanwhile, the practice of caring for the sick fell largely to religious figures and institutions, and

The restored Roman baths in Bath, Somerset. Built in AD *75, they were reopened in 1897 and again in 2011.*

the concept of disease intertwined with ideas of divine punishment and spiritual healing.

Progress was initially much more pronounced in the Islamic world. It was here that the important work of preserving ancient scholarly works took place, preventing much of the knowledge-loss experience in early medieval Europe. Medieval Islamic physicians made significant contributions to medicine, particularly in the fields of anatomy, surgery and drug treatments. The great Persian polymath Abu ibn Sina (sometimes known by his Latinised name Avicenna) is still regarded as one of history's greatest physicians, and is sometimes described as the father of early modern medicine. His five-volume medical encyclopaedia, *The Canon of Medicine*, would go on to be the primary medical textbook used in the West, right up until the 18th century.

The work of medieval Islamic scientists influenced scholars in Europe and became a central pillar of Renaissance medical thought, alongside Greco-Roman works. Between the 16th and 18th centuries, the rate of medical advancements increased and medical education improved, as university curriculums became more rigorous and scholars began to challenge classical thought, giving way to new ideas and discoveries.

A 13th-century artwork depicting an Islamic surgeon at work.

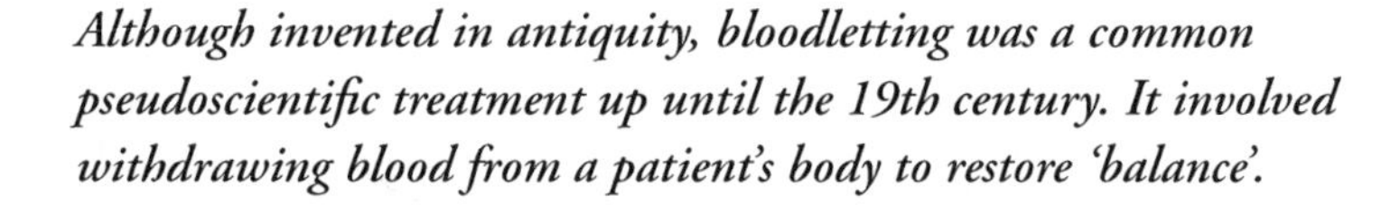
Although invented in antiquity, bloodletting was a common pseudoscientific treatment up until the 19th century. It involved withdrawing blood from a patient's body to restore 'balance'.

Scientific medicine as we now recognize it began to develop from the 19th century onwards. During this period, advances in physicians' understanding of the causes of disease, alongside enhanced public health interventions, helped to yield widespread health improvements. In 1849, an English physician named John Snow used statistical methods to trace the source of an outbreak of cholera in London to one particular contaminated water pump; the pump handle was removed and the epidemic subsequently subsided. Snow is often credited as the father of epidemiology: the study of health and disease among groups of people, on a collective rather than an individual basis.

Did You Know?

Hygiene is now a fundamental part of modern medicine, but that wasn't always the case. A Hungarian physician named Ignaz Semmelweis was the first to note the link between maternal mortality and hand washing in the 1840s, but his calls for medical hygiene were largely ignored until the late 19th century.

In the 20th and 21st centuries, medical treatments continued to advance and mortality rates decline throughout most of the world. The invention of antibiotics in the early 20th century and more recently the development of vaccine technology have saved countless lives and dramatically improved life expectancy around the world. In addition to improved drugs, modern science has also seen the identification of previously unknown factors in human health, including hormones, vitamins, genetics and lifestyle factors. We are still progressing in the study of human health and medicine, and advances in genetics, technology and microbiology all have the potential to revolutionize medical treatments in the years to come.

The physician John Snow, photographed in 1856.

Medicine did not emerge out of nowhere; it has been part of human life since the dawn of civilization. Over time, medicine has branched off into numerous subdisciplines, each with its own extensive history of scholarship, but we still owe a great deal to ancient civilizations. Living in a time long before microscopes, stethoscopes and thermometers, societies eked out information about the human body and ways to care for it through sheer ingenuity and observation. While they weren't always correct in their deductions, their quest for understanding helped to advance human knowledge and to give us better control over our health and well-being.

CHAPTER 3
PHILOSOPHY

'Wonder is the beginning of wisdom.'
– SOCRATES, GREEK PHILOSOPHER

The word 'philosophy' literally means 'love of wisdom'. Scholars of philosophy look to build knowledge in a structured way, often asking fundamental questions about the nature of existence, humanity and knowledge itself. As such, philosophy covers a broad range of topics that sit outside of the empirical sciences. Some of the cornerstones of modern science are the use of controlled experimentation, the search for solid facts and the development of testable theories. Meanwhile, philosophy uses techniques such as rational debate, critical reasoning and the use of questions to acquire knowledge. So where a modern scientist might ask, 'what are the specific chemicals found inside DNA?', a philosopher might ask, 'what does it mean to be alive?'

But science and philosophy weren't always considered separate disciplines. In fact, from the 3rd millennium BC up until the 19th century, much of what we now term 'science' was referred to in the West as 'natural philosophy'. Western philosophy has its roots in Ancient Greece, starting in around 600 BC, and this philosophical tradition has shaped the story of science, giving rise to scientific ideas and approaches that influenced scientists for centuries to come. The ancient philosophers helped to develop many interesting areas of scientific enquiry, including mathematics and medicine, but perhaps their greatest contribution to the history of science was the rational way that they developed ideas and knowledge. In this chapter, we will be exploring the ways in which philosophical thought intersected with science and informed the development of the modern scientific method: the system by which scientists answer questions about the world around us.

Over time, Western philosophy has branched off in many different directions, eventually establishing itself as one of the humanities – fields of study that relate to human culture, such as literature, history and languages. However, the Ancient Greek philosophical tradition laid the foundations for science to grow into an entirely separate entity, one that would go on to shape the course of human history.

c. 6th–5th CENTURIES BC	*Pre-Socratic philosophers develop notion of the 'kosmos'*
c. 5th CENTURY BC	*Socratic philosophers develop 'Socratic method' of thought*
399 BC	*Socrates put to death for 'corrupting the minds of the youth of Athens'*
c. 387 BC	*Plato sets up the Academy: an early organized academic institution*
4th CENTURY BC	*Aristotle establishes principles of logic through deductive reasoning*
965	*Ibn al-Haytham establishes importance of scientific experimentation*
1620	*Francis Bacon establishes founding principles of scientific method*
1644	*René Descartes questions reliability of the senses*
1664	*Galileo Galilei champions importance of independent confirmation*
1687	*Isaac Newton develops modern scientific method*

ANCIENT PHILOSOPHY

The Ancient Greek philosophers helped to develop the foundations of a number of core scientific ideas. A branch of wider philosophical thought, natural philosophy was broadly concerned with the study of the natural and physical world, and incorporated a broad range of scientific disciplines, including rudimentary physics, chemistry and biology. But perhaps the greatest legacy of the Ancient Greek philosophers is not *what* they thought, but rather *how* they thought.

It's hard to overstate the significance of the philosophical method in the history of science. From prehistoric times, people were seeking out knowledge and using it to their advantage, often passing information down from generation to generation. But Ancient Greece was home to a revolution in Western thought that saw the systematic ordering of knowledge, and the logical development of that knowledge into formal theories. The Greeks weren't the first people to think in this way – this approach also has roots in Ancient India and China – but they did help to popularize the use of logic and reason to generate inquiry and develop understanding.

Logical thought was able to thrive in Ancient Greece partly thanks to the Greeks' somewhat more secular philosophical tradition. The belief in gods and goddesses who shaped the human experience was very much widespread in Ancient Greece, but most philosophers also believed that patterns and rules existed in nature and could be understood through study. This separation between faith and fact prompted the early philosophers to search for fundamental laws of nature and reality, instead of consistently turning to a supernatural entity for explanations.

The pre-Socratics were some of the earliest philosophers to think in this way. They developed the idea of the 'kosmos': a universe ordered by rules that could be understood using rational inquiry. The pre-Socratics asked big questions, such as 'what is the universe made of?' and 'where did the world come from?', and they used reason and critical thinking to try to arrive at answers to these questions.

Who's Who – Thales of Miletus

Born around the year 624 BC, Thales of Miletus was one of the most renowned pre-Socratic philosophers, and is lauded by some as the father of Western philosophy.

One of Greece's 'Seven Sages' – the esteemed wise men of antiquity – Thales was accomplished in geometry and astronomy. It is said that he once calculated the size of Egypt's great pyramids just by measuring their shadows, and that he accurately predicted a solar eclipse that took place on 28 May 585 BC.

But perhaps Thales' greatest contribution to science is the way in which he approached the study of natural phenomena. Rather than looking to gods and goddesses as the cause of natural events, Thales looked for non-supernatural explanations. This approach was revolutionary – anthropomorphic gods were at the centre of Greek thought at this time, and religious belief was often intimately intertwined with protoscientific enquiry. By relying on reason rather than religious belief, Thales helped to establish a new way of approaching problems and laid the foundations for the growth of science.

Socrates drew on these approaches when he developed his 'Socratic method': a form of cooperative debate in which individuals asked questions and argued points in order to stimulate critical thinking and arrive at logical conclusions. In 399 BC, Socrates was put to death for impiety and for 'corrupting the minds of the youth of Athens' with his revolutionary ideas. But this was by no means the end for Ancient Greek philosophy.

Socrates was succeeded by his student Plato, who went on to set up one of the world's earliest organized academic institutions. Then Plato was succeeded by his student Aristotle, who developed the scientific foundations of philosophy even further. Unlike Plato and many of the other philosophers who came before him, Aristotle emphasised the importance of empirical observation and experimentation as pathways to knowledge. He argued that by observing patterns and phenomena in the world, we could build up a bigger picture of how things work.

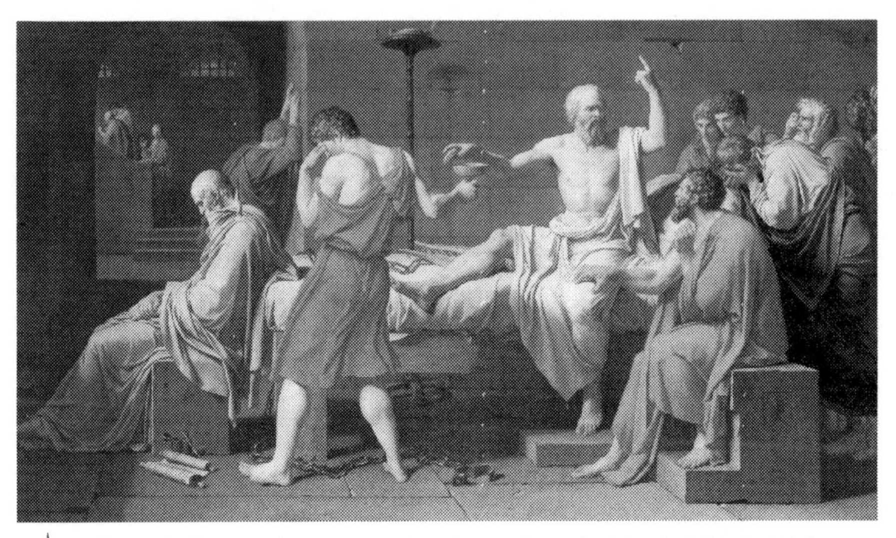

The Death of Socrates, *by Jacques-Louis David, 1787, depicts the philosopher's forced suicide by hemlock.*

Plato's Academy

Somewhere around 387 BC, the philosopher Plato founded his academy, a key event in the history of Western education. Plato's Akademia was one of the first known organized institutions of higher learning, and could be considered an ancestor to modern universities.

Situated in an olive grove just outside the city walls of ancient Athens, the academy was a place for knowledge-hungry students to sit and discuss philosophical teachings and listen to lectures from Plato and his contemporaries.

But the academy was not open to just anyone. It is thought that a sign above the school's entrance read something to the effect of: 'let no one ignorant of geometry enter here'. The academy encouraged the spread of reason and observation, and developed the principle of dedicated institutions for higher learning.

Aristotle is also widely considered to be the father of logic. He developed a form of deductive reasoning in which two or more true statements can be used to arrive at a logical conclusion. This approach to forming knowledge marked a departure from the more abstract, theoretical approach taken by his predecessor Plato. Aristotle wasn't always fully methodological in his approach and, like his contemporaries, would sometimes rely on untested knowledge in his thinking. But nonetheless, Aristotle's focus on empirical evidence and logical reasoning was revolutionary, earning him a place in the history books as one of the first *true* scientists.

Did You Know?
The most famous example of Aristotle's deductive reasoning is:

A: All men are mortal.
B: Socrates is a man.
C: Thus, Socrates is mortal.

Using these methods of logical reasoning, thinkers can use existing knowledge to develop new knowledge. Of course, when the accepted premises in points A and B are wrong, the conclusion, C, will also be false.

The Ancient Greek philosophers were by no means the first people ever to think about the world around them, but their adoption of systematic methods of study and their attempts to formalise knowledge helped to shape the development of science for centuries to come. Indeed, while philosophy and science would ultimately become separate disciplines, the legacy of the Greek philosophers continues to inform scientific thought even today.

Did You Know?
The difference in thinking between Aristotle's empirical approach and Plato's more abstract ideas was famously represented in

Raphael's 16th-century painting The School of Athens, *in which Plato is depicted pointing upwards to the heavens while Aristotle gestures downwards towards the Earth, symbolizing his more grounded approach.*

THE ROAD TO MODERN PHILOSOPHY

The Ancient Greek philosophers might never have exerted their influence on scientific thought without the help of Islamic scholars. In the early Middle Ages, many of the classical Islamic works were unknown in the West, and philosophical thought had become closely tied to theology and religion. But far from being a global 'dark age' for science and reason, in the Islamic world the Middle Ages saw dramatic intellectual and academic progress. This period – known as the Islamic Golden Age – lasted from the 7th to the 10th centuries; during which time scholars rescued countless Greek texts from obscurity, translating them into Arabic and preserving the knowledge for future generations.

In the Islamic world, philosophical ideas about how we acquire knowledge were built upon, creating the early foundations for the modern scientific method: the process by which scientists study and uncover information about the world around us. Born in Iraq around AD 965, the mathematician and astronomer Ibn al-Haytham (see page 70) was instrumental in the development of this method. Al-Haytham tested his scientific theories through systematic experimentation: a cornerstone of scientific enquiry that had been largely lacking in the Ancient Greek approach. For Islamic scientists such as Al-Haytham, experimentation came to play a much greater role than it had done in the classical world.

During the Renaissance, European thinkers returned to the works of the Ancient Greeks, and themselves began to build on the philosophical method laid out by Aristotle and his contemporaries. Galileo Galilei, an Italian polymath born in 1664, is remembered for his monumental contributions to astronomy. But Galileo also helped to develop the early scientific method by calling for scholars to perform their own tests in order to prove experimental theories. Galileo argued that independent confirmation was essential to the development of scientific theories – it wouldn't do to just to rely on the observations and ideas set out by other thinkers – theories must be tested.

Around the same time, the British philosopher and statesman Francis Bacon publicized his own vision of the scientific method, articulating many of the principles as we know them today. Bacon recommended that the state support scientific enquiry, and that science seek to improve the welfare of humanity through progress in technology and other applied sciences. He also argued that useful knowledge could only be gained through first-hand testing of theories – Bacon agreed with Galileo that it wasn't enough to trust the words of ancient scholars. For him, scientific knowledge depended on experimentation.

Applied science refers to putting scientific findings to practical use, often through technology, engineering and inventions.

Meanwhile, writing from the Netherlands, the French philosopher and mathematician René Descartes was formulating an independent theory of knowledge. As a philosopher, Descartes was interested in the question: 'how do we know what we know?'. This question arises from a philosophical field known as 'epistemology', which is concerned with the nature of knowledge itself. Descartes argued that Aristotle's emphasis on observation as a path to knowledge was flawed, because we cannot always trust our own senses. Instead, Descartes suggested that systematic doubt and rational questioning of one's own senses is essential to developing accurate knowledge.

Ultimately, it was with Isaac Newton that the scientific method reached its maturity. Published in 1687, his treatise *Principia* brought together the various rules for the generation of accurate scientific knowledge. Like Bacon and Galileo before him, Newton emphasised the importance of experimentation as a route to knowledge. He held that knowledge is best acquired by observing the world, developing theories to explain what is seen, and then carrying out tests to see if the theory is correct.

Galileo Galilei argued for the importance of the first-hand testing of experimental scientific theories.

Epistemology is a branch of philosophy concerned with the theory of knowledge and how we acquire it.

By the close of the 17th century, the modern scientific method had become well established, and although it has been refined over the years by many great scientists and philosophers, the basic tenets of observation, theorizing and experimentation have held true.

The scientific method is typically applied as follows: the presentation of a hypothesis based on observable facts, the testing of that hypothesis under experimental conditions, and the refinement and development of the hypothesis in line with the results of experiments.

Science has its origins in philosophy, and philosophers have been instrumental in developing the core principles of scientific enquiry. While science and philosophy are now two very separate disciplines, we still see philosophy intersect with science in interesting ways. The field of theoretical physics sometimes pushes the boundaries of scientific enquiry into the philosophical realm. For instance, when we consider the question of what the concept of time meant before the Big Bang, we're asking both a scientific and a philosophical question. Meanwhile, ethics and morality frequently feature in scientific discourse, and navigating those issues often warrants both a scientific and philosophical approach. Ultimately, scientific enquiry and its associated advancements affect the world around us in myriad ways. Questions about what areas of science we prioritise, how we conduct experiments and how we manage the potential consequences of scientific progress all draw on both scientific and philosophical schools of thought. Science is an inherently human pursuit, and therefore philosophy will always have a role to play.

PART TWO

THE POST-CLASSICAL ERA

5TH–15TH CENTURIES

THE POST-CLASSICAL ERA – which roughly corresponds to the Middle Ages – is often thought of as a dark time in human history, a time when science gave way to ignorance and superstition. There is some truth in this; after the decline of the Roman Empire in the 5th century, scientific progress slowed and much knowledge was lost. But this period also saw genuine scientific innovation and advancement. This was particularly true in the Islamic world – here, classical scientific knowledge from Greece and Rome was preserved and eventually transported to Western Europe through booming trade networks and war. In Song dynasty China, a spike in industry coincided with a spate of groundbreaking engineering projects and inventions, including moveable type printing, gunpowder and paper money.

Scientific endeavour in Eurasia was set back by the outbreak of the Black Death in the 14th century – a fearsome pandemic which killed millions of people and decimated the populations of Europe and the Middle East. But the end of the post-classical period saw an upsurge in cultural activity, culminating in the European Renaissance from the 15th century. To dismiss the post-classical era as a quiet, backwards period between two eras of enlightenment is to overlook the many scientific and technological advancements made during this portion of human history. Far from its characterisation as a 'dark age', the post-classical era saw thinkers and innovators continue the pursuit of scientific truth and enlightenment.

CHAPTER 4
GEOGRAPHY

'Geography is an earthly subject, but a heavenly science.'
– EDMUND BURKE, IRISH POLITICIAN AND PHILOSOPHER

Geography – which translates from the Greek for 'earth description' – is the study of places, their physical features and the interactions between people and the environment. Although the word originates in Ancient Greece, people have been practising geography in some form all over the world for centuries. Geography flourished in Ancient Greece, but it became its own profession in Ancient Rome. In the post-classical era, the study of geography made great strides in the Islamic world, supported by Middle Eastern scholars' profound knowledge of mathematics, eventually returning to Europe during the Renaissance. These events set the stage for the great explosion in geographic study that took place in the 18th and 19th centuries.

Geography has always been valuable because of its practical uses – understanding the layout of regions is important for trade and warfare – but it gives us much more than that. Geography provides another perspective on the story of human life; it helps us to understand the environment and its impact on us, and gives structure and reason to our surroundings.

C. 9th CENTURY BC	*Earliest evidence of maps produced in Babylon*
C. 3rd CENTURY BC	*Eratosthenes estimates circumference of the Earth*
1037	*Al-Biruni theorizes existence of the Americas*
1154	*Muhammad al-Idrisi produces influential world map: the Tabula Rogeriana*

12th–13th CENTURIES	*Christian crusades extend European geographical knowledge*
c. 1300	*Marco Polo publishes account of his travels in the Far East*
c. 15th CENTURY	*Dawn of the Age of Discovery*
1492	*Christopher Columbus lands in the Americas*
1522	*Ferdinand Magellan's crew circumnavigate the globe*
1830	*Royal Geographical Society is established in England*

THE ROOTS OF GEOGRAPHY

Long before the scientific discipline of geography existed, human beings were exploring and charting the environment around them. The earliest maps are thought to date back to 9th-century BC Babylon, and the oldest surviving world map also originates in this region. Produced at some point between the 7th and 5th century BC, the Babylonian World Map depicts Babylon at the centre of the world, surrounded by a salt sea and islands.

While geography existed in practice for many years prior to this, it was in the Mediterranean during the Hellenistic Period that it gained its name. By the 4th century BC, the Ancient Greeks had an impressive geographical knowledge; they divided the world into three continents – Europe, Asia and Africa – and scholars widely accepted that the Earth was a sphere rather than a disk. The father of geography is generally thought to be the scientific writer, mathematician and astronomer Eratosthenes. Born in Greece in 276 BC, Eratosthenes spent most of his life in Egypt, where he worked as chief librarian at the Library of Alexandria. It was here that he made the first accurate calculation of the circumference of the Earth, and published *Geographika*, a comprehensive work on the origins, measurements and layout of the Earth. Eratosthenes didn't get everything right – he believed that the world was divided into five climate zones, just two of them inhabitable – but his work contained

 *This clay tablet, the oldest surviving world map, puts Babylon at the centre of the world.*

a wealth of accurate geographical knowledge that laid the foundations for the development of geography as a discipline.

Did You Know?

You may have heard that Columbus discovered that the world was round, but this is a common misconception. The Ancient Greeks were widely aware that the Earth was not flat, and the mathematician Pythagoras is said to have proposed a spherical Earth in the 6th century BC.

In Ancient Rome, geography was seen as a useful tool in the building of empire, and the discipline was valued by statesmen and military leaders. Accurate maps and geographical measurements allowed the Romans to create their vast road transportation system. Many Roman roads still survive today, and are admired for their efficiency – a trait that depended on the work of expert geographical surveyors. The geographer Strabo, writing in the 1st century BC, and the polymath Ptolemy, writing in the 2nd century, extended the scope of geography by compiling descriptions of countries, local resources and customs. Ptolemy's geographical treatise and atlas, *Geographia*, would go on to influence geographers in both the medieval Islamic world and in Renaissance Europe.

Did You Know?

Eratosthenes made his calculation of the Earth's circumference by measuring the angle of the Sun's rays in two separate places, around 500 miles apart. From this figure, he was able to extrapolate that a spherical Earth would have to be between 25,000 and 28,500 miles in circumference. The true figure is just under 25,000 miles, so Eratosthenes' calculations were remarkably close.

In the centuries following the fall of Rome, various societies' geographical knowledge was aided by the growth of the Silk Road: a network of trade routes that effectively connected China with the Mediterranean. Geographical knowledge was spread through the Silk Road traders, whose stories of far-off lands helped nations to create a more accurate picture of the world outside their borders.

Did You Know?

By the 2nd and 3rd centuries, trade along the Silk Road was booming, but it wasn't just goods that were transported from one continent to another; trade routes also helped to transport ideas such as religion, art and science across great distances.

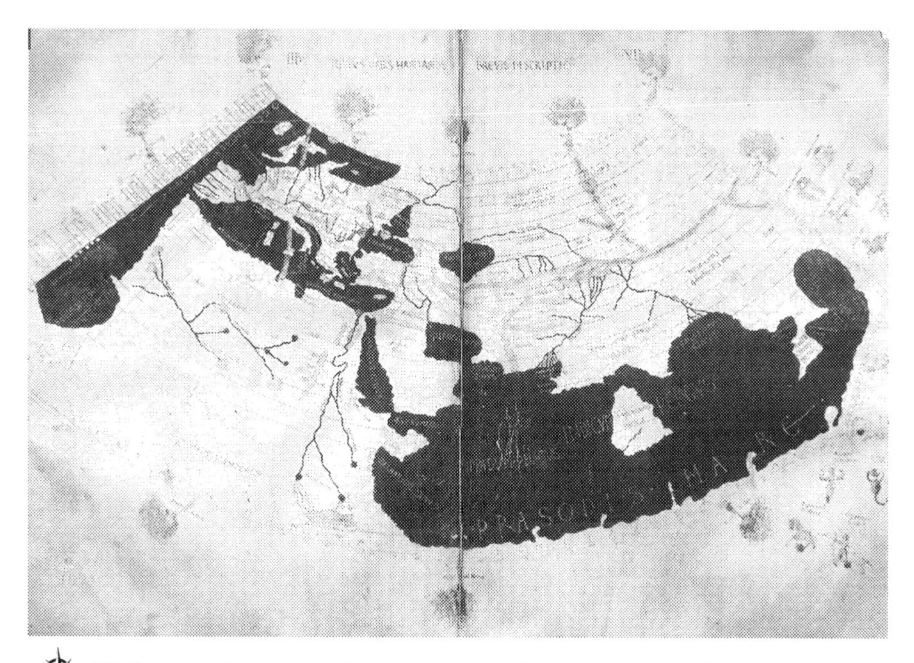

The Ptolemy world map depicts the world known to Hellenistic society in the 2nd century.

Over the course of the ancient era, world maps transformed and grew, as civilizations separated by thousands of miles became aware of one another for the first time. Technology, politics and trade also prompted the transition, and gradually in the early centuries of the first millennium, the world began to open up.

The Islamic Golden Age

Beginning during the Umayyad dynasty in AD 692, the Islamic Golden Age refers to an era of Islamic conquest that coincided with a period of political stability and cultural and scientific growth. In 762, the capital of the Islamic empire was moved to the newly founded city of Baghdad, in modern-day Iraq. Here, the ruler of the empire, Caliph al-Ma'mun, established the Baghdad House of Wisdom – a library and one of the largest centres of learning in the world. The House of Wisdom became a hub for scholars, philosophers and scientists, and it was here that the 'translation movement' prompted the preservation of many important texts.

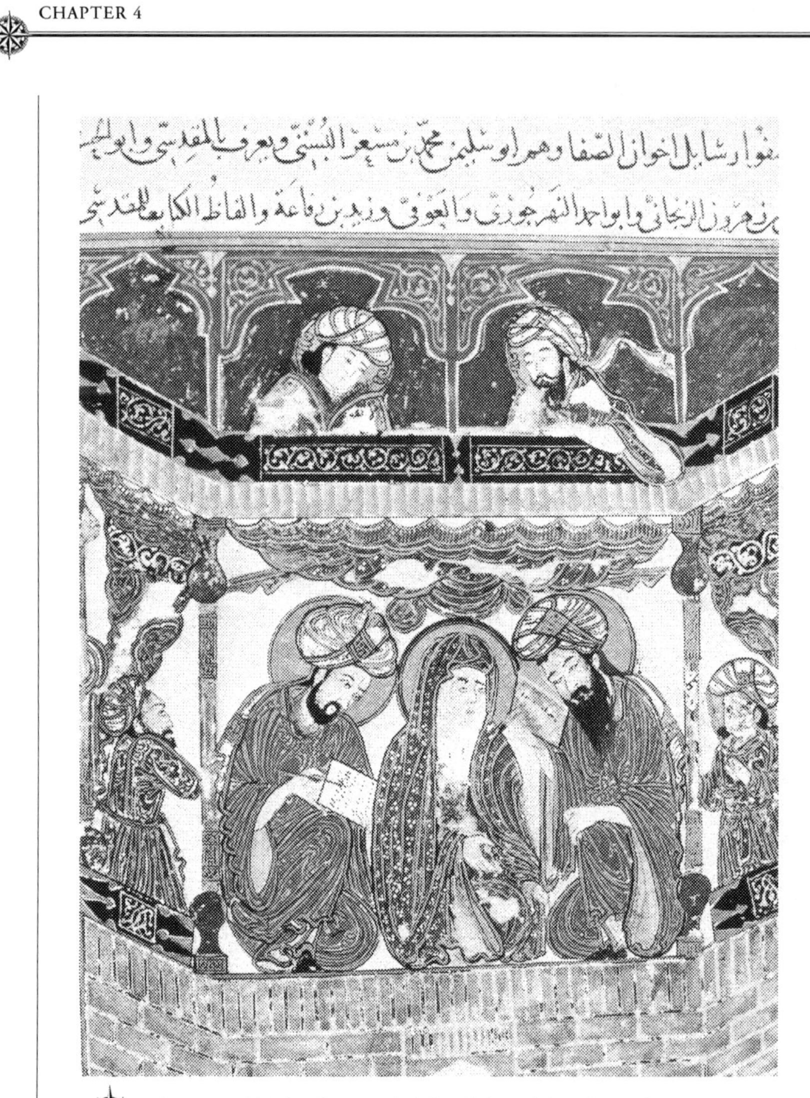

The House of Wisdom became a hub for scholars, philosophers and scientists.

It wasn't just the Ancient Greeks who influenced Islamic scholars during the Golden Age. Ideas such as the Hindu-Arabic numeral system (see page 19), which came from India, and printmaking, which originated in China, were also adopted and adapted in the Islamic world.

But the Islamic Golden Age wasn't just about translating and passing on existing knowledge. Islamic scholars also built on the knowledge that they gathered from Greece and elsewhere, and made progress in the fields of mathematics, astronomy and medicine. Scholars such as Al-Khwarizmi in mathematics, Ibn al-Haytham in astronomy and physics, and Ibn Sina in medicine, are remembered among the most accomplished thinkers in scientific history.

The Islamic Golden Age shaped the character of the empire itself, but it also impacted on the rest of the world. Because the empire's capital, Baghdad, was strategically positioned between Europe and Asia, it became a crossroads for both trade and ideas. In this way, the Islamic world acted as a bridge of knowledge between East and West – this effect was particularly pronounced in Europe, where many scientific ideas passed first through the Islamic empire before arriving in the West.

The Islamic Golden Age is traditionally held to have come to an end in 1258, when Baghdad fell to a Mongol invasion. The House of Wisdom was destroyed by the Mongols and the ruling caliph was executed, but the Golden Age left behind a legacy, particularly in Europe. Although there isn't consensus on whether to call the Islamic Golden Age a 'scientific revolution', it's clear that the emphasis on learning and knowledge spurred substantial scientific progress, which in turn supported the growth of ideas and enquiry in Europe. There's no doubt that the sciences owe a debt to the Islamic world, and to the thinkers of the Islamic Golden Age.

POST-CLASSICAL GEOGRAPHY

Geographical progress slowed in Europe during the Middle Ages, but the discipline continued to thrive in the Islamic world. In the medieval period, Islamic scholars began the great translation movement, which brought a great deal of new geographical knowledge into the Islamic world and allowed Arabic scholars to build on existing scholarship. These texts were stored at the House of Wisdom – a centre for academic learning located in Baghdad. Islamic geographers, such as the 9th-century Persian scholar Al-Khwarizmi, revised and built on these texts, correcting errors and adding new knowledge.

Did You Know?
The translation movement took place during the Islamic Golden Age, and involved translating ancient texts – primarily from China, India, Greece and Rome – into Arabic.

In the 11th century, the Iranian polymath Al-Biruni is thought to have become one of the first scholars to theorize the existence of the Americas. Based on his calculations of Earth's circumference and known landmasses, and using his knowledge of geological processes, Al-Biruni deduced that another large landmass probably existed somewhere in the ocean between Europe and Asia.

In the 12th century, the North African geographer Muhammad al-Idrisi produced the *Tabula Rogeriana*, a world map complete with descriptions of the physical, cultural and political characteristics of various regions. Al-Idrisi spent 15 years compiling his map, under the commission of King Roger II of Sicily, and it remained the most accurate depiction of world geography for many years afterwards. One of the most striking things about Al-Idrisi's work is the level of knowledge he ascertained about far-off places. The *Tabula Rogeriana* contains descriptions of the Chinese silk trade and of various islands in the remote North Sea. At the time, Al-Idrisi's work must have seemed like an unprecedented insight into mysterious lands that others could only imagine.

In medieval Europe, geographical progress staggered for some centuries before the Renaissance. However, war and trade both continued to expand scholars' knowledge of the world beyond Europe and piqued interest in exploration and travel. The travels of the Italian merchant Marco Polo helped to fuel Europeans' curiosity about other regions. Polo's story of travelling from Venice to the Far East, a journey of some 5,600 miles

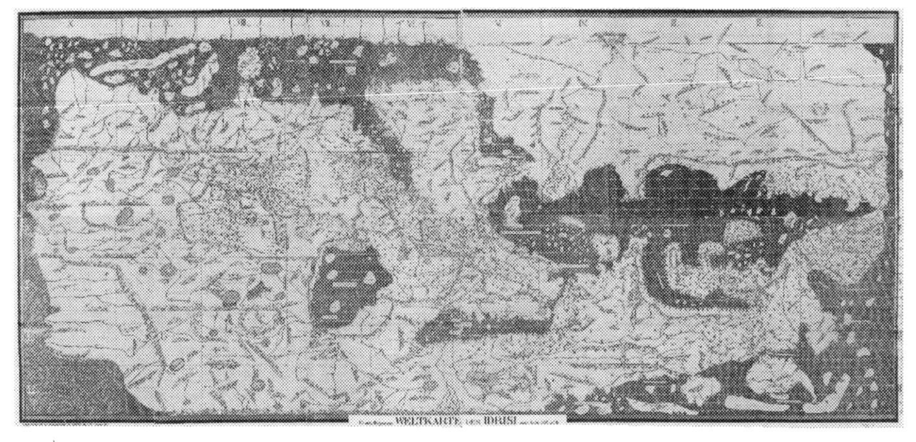

Arab geographer Muhammad al-Idrisi produced an influential world map known as the Tabula Rogeriana.

along the Silk Road, was popularized in a book released around the year 1300, which went on to inspire other European explorers, including Christopher Columbus. Meanwhile, the Christian crusades that took place between the 12th and 13th centuries helped to increase European knowledge of the Islamic world.

Towards the end of the post-classical era, in the 15th century, European nations began conducting long sea voyages in an effort to explore uncharted areas of the world in search of gold, silver and spices. Known as the Age of Discovery, this period lasted up until the end of the 18th century, and saw the discovery of many regions previously unknown to Europeans, including the Americas. For the native inhabitants of these lands, the arrival of the Europeans would lead to exploitation and oppression. Colonial policies in Europe decimated native populations and triggered the emergence of the Atlantic slave trade, ultimately creating a level of economic and political dominance in Europe that can still be seen to this day.

THE ROAD TO MODERN GEOGRAPHY

From the 15th century onwards, geography experienced a major revival in Europe, driven by the twin goals of commerce and conquest. As Europe's imperialist ambitions grew, so too did the importance of geography. A series of sea voyages in the 15th and 16th centuries – including Christopher Columbus' accidental discovery of the Americas and Ferdinand Magellan's circumnavigation of the globe – helped to highlight the practical significance of the discipline. Information about the physical world became immensely valuable to governments looking to expand beyond their own borders, and geography soon transformed into a booming business. Colonial fervour continued well into the 19th century, as European countries engaged in fierce competition over colonized lands, and all the while geography grew in status.

This trend ultimately triggered the 'professionalisation' of geography. In the 19th century, professional societies began to spring up around the world, including the Royal Geographical Society in England and the National Geographic Society in the USA. To cement the position of geography as a respected academic discipline, learned societies successfully petitioned for universities to begin teaching the subject.

Between the late 19th and early 20th centuries, European countries rushed to colonize countries in the African continent. This imperialist venture came to be known as the 'scramble for Africa'.

Who's Who – Alexander von Humboldt

Born in Prussia in 1769, Alexander von Humboldt is an integral figure in the history of geography, but his influence extends far beyond that. A true polymath, Humboldt also left his mark more broadly on the fields of naturalism, philosophy and scientific endeavour.

From a young age, Humboldt was determined to become a scientific explorer. In 1799, after having undertaken several trips within Europe, he embarked on a voyage to the Americas. He would spend the next five years travelling around the continent, passing through Venezuela, Cuba, the Andes, Mexico and the USA. When he returned from his voyage, Humboldt wrote about his experiences in the New World, establishing himself as an international name within the scientific community. His meticulous documentation of plant and animal species and comprehensive descriptions of the regions he visited helped to expand the sciences of geography and naturalism, and brought an evocative image of Latin America to the many readers of his bestselling books.

Humboldt was a prolific writer, publishing 23 volumes of travel writing in total, but perhaps his most significant text was his final one, Cosmos. Published in five volumes, beginning in 1845, Cosmos was Humboldt's attempt to bring together all of his scientific knowledge in one grand theory of the universe. The book was well received, and Humboldt's writings would go on to influence a number of great minds, including Charles Darwin himself.

A staunch abolitionist and early environmentalist, Humboldt was in many ways a man ahead of his time. He was one of the first scientists to propose that various continents were once connected, and was the first to discuss the concept of human-led climate change.

Once one of the most famous men in all of Europe in his era, Alexander von Humboldt's name is no longer well known. Nonetheless, his contributions to geography are unparalleled, and he deserves recognition as one of the true icons of scientific history.

While the 19th century was a period of establishment and professionalisation for geography, the 20th century saw the broadening of the subject into separate subdisciplines, which can be roughly grouped into two core fields of study: physical and human geography. Physical geography is focused on environmental processes and systems, such as ecosystems, climate and the atmosphere; while human geography focuses on the relationship between people and places, looking at cultural phenomena such as political and social structures. As such, geography is as relevant now as it has ever been. Although we no longer have large blank spots on the map, there are still plenty of questions left for geographers to answer. From population change and climate to food security and biodiversity, geography remains vital to confronting the challenges of our age.

The evolution of geographic thought has contributed to changes in the way that we as human beings see ourselves. The first world maps reflected their makers' limited understanding of their place in the world – to them, they were citizens of Babylon, and Babylon was the world. But as geography advanced, so too did different civilizations' knowledge and understanding of one another. In the 21st century, we remain citizens of our various nations, but we also know how much else is out there. And that makes us part of a much bigger story. Thanks in part to the study of geography, we have become not just citizens of our various nations, but also citizens of the world.

CHAPTER 5
OPTICS

'[Optics] is the flower of the whole of philosophy and through it, and not without it, can the other sciences be known.'
– ROGER BACON, MEDIEVAL PHILOSOPHER AND SCIENTIST

The science of optics is concerned with both the behaviour of light and the study of sight. We can understand optics as being broadly divided into two main subdisciplines: physical and geometrical optics. Physical optics is concerned with the nature of light itself as a wave, while geometrical optics considers the behaviour of light as a beam and how that beam can be manipulated using our eyes, lenses or other instruments that affect light.

Human beings have always had an important relationship with light. The symbolism of light as a power for good appears in many religious texts; and on a practical level, being able to exploit light has helped humans to become a dominant species. Our understanding of light has increased over the centuries, with scholars all over the world contributing to optical science. These days, optics is at the core of modern science; from the telescopes that we use to explore the universe to the microscopes that help us diagnose disease and develop drugs, an understanding of optics and optical instruments is crucial to scientific progress. There's still plenty that we don't understand – some optical phenomena intersect with the field of quantum mechanics – but the knowledge that we've built up over the centuries has helped to advance science in myriad ways. This is the story of how we got from the earliest theories of light and optical tools to where we are now, and how we came to understand and control one of the most powerful tools known to humankind.

c. 8th CENTURY BC	*Early magnifying lenses developed in modern-day Iraq*
c. 6th CENTURY BC	*Pythagoras proposes early theory of vision*
984	*Ibn Sahl publishes theory of refraction*
1021	*Ibn al-Haytham develops theory of eye anatomy and function*
1690	*Christian Huygens proposes that light is a wave*
1704	*Isaac Newton suggests light is made up of particles*
1865	*James Clerk Maxwell proposes theory of electromagnetism*
1900	*Discovery of gamma rays completes electromagnetic spectrum*
1905	*Albert Einstein establishes theory of wave-particle duality*

THE ROOTS OF OPTICS

Long before we knew anything about the physical properties of light, human beings were trying to manipulate it. Magnifying lenses from the 8th century BC have been unearthed in modern-day Iraq, suggesting that ancient Mesopotamian civilizations were aware of how to bend and refract rays of light, although they had no notion of the physics behind this effect.

In Ancient Greece, philosophers developed some of the earliest theories on the nature of light and sight. In the 6th century BC, Pythagoras proposed that the eye projects light rays on to an object, allowing us to see that object; some years afterwards, Democritus theorized that the source of vision is not in the eye, but in the objects that we behold. He argued that objects project a stream of images which then enter the eye, allowing us to see whatever we're looking at. In the 5th century BC, Empedocles combined these two conflicting notions to hypothesise that the eye does indeed project light, but that this interacts with external sources such as the Sun to create sight. Of course, these theories now seem completely ridiculous, but the Ancient Greeks had little in the way of scientific evidence to help them understand how light behaves, nor did they have enough anatomical understanding of the eye to understand its role in vision.

Around the dawn of the 3rd century BC, the mathematician Euclid laid the foundations for geometrical optics when he published a series of rules or axioms about how light behaves. While Euclid maintained the erroneous idea that vision is formed by light being projected outwards from the eye, he did also make a number of accurate observations. Notably, his work put forward the notion that light travels in straight lines, meaning that it can be studied using geometrical methods. He also suggested that light can be manipulated, and reflected back in a different direction. In the 1st century, the Alexandrian mathematician Hero established the rule that light always takes the shortest route, and in the 2nd century the Roman polymath Ptolemy observed that light can be refracted as well as reflected.

Did You Know?

Centuries before the invention of cameras, another device existed: the camera obscura. From the Latin meaning 'dark chamber', the camera obscura is a box with the inside painted black. A small hole on one side lets in light, which then creates an inverted image (upside down and back to front, although with colour and perspective preserved) of the outside scene on the back wall of the box.

The same effect can be created in a darkened room. If a small pinhole of light is let in from a window, an inverted image of the scene outside will be projected on to the opposite wall. This happens because the light travels in straight lines, becoming focused through the tiny hole. After passing through the hole, the light beams cross over one another, leaving the inverted image that they create on the back wall. The human eye works in a similar way to a camera obscura, with the pupil acting as the tiny hole and the retina as the opposite wall where the image is formed.

Camera obscuras have existed in some form since ancient times, but the first written evidence heralds from 4th-century China. A description of the camera obscura is attributed to the Chinese philosopher Mozi, centuries before any such description would appear elsewhere. During the 10th century, the Chinese scientist Shen Kuo used the camera obscura in his optical experiments, describing its mechanism in detail. Around the same time, the technology appeared in the Islamic world, where it was used by the Iraqi scientist Ibn al-Haytham to carry out his studies into optics. Through the 17th and 18th centuries, camera obscuras continued to be used by artists, scientists and the general public alike. However, as newer technology emerged – such as the magic lantern — camera obscuras declined in popularity. These days, the devices are not widely in use; however, anyone is able make their own with little more than a knife, a box and some tape.

By the end of antiquity, the field of optics had begun to grow. Scientists were learning more about how light beams behave and interact with our vision, but there was still much about the behaviour of light that remained a mystery. In addition, scientists were still no closer to understanding the physical nature of light, what it's made of and the physics of how it behaves. To understand how the story of optics progressed in the post-classical age, we must turn to the Islamic world.

POST-CLASSICAL OPTICS

Like many other scientific disciplines, the field of optics experienced a period of resurgence during the Islamic Golden Age, which lasted from the 8th to the 15th centuries. It was during this time that great Islamic scholars and mathematicians advanced many aspects of optics, and established some of the scientific theories and principles we still use today.

Ibn Sahl was a Persian mathematician who practised in Baghdad in the 10th century. Intrigued by the optical theories proposed by Ptolemy centuries before, at some point around the year 984 Ibn Sahl published his own theory of refraction, which built on Greco-Roman ideas and quantified the way in which light changes direction when it moves between mediums. This was a major achievement – it would be another 600 years before the same principle was described in the West. Ibn Sahl's pioneering work is thought to have influenced the scholar who would later become known as 'the father of modern optics', Ibn al-Haytham.

A polymath with expertise in astronomy, mathematics and physics, Al-Haytham published a large treatise in the 11th century, *The Book of Optics*, which set forward a number of principles on light and vision that would revolutionize the field. Al-Haytham recognized that vision depends on light entering the eye, not on the eye projecting light outwards as Euclid and other Greco-Roman thinkers had argued. Through a series of experiments using lenses, mirrors and other optical devices, Al-Haytham supported Euclid's theory that light travels in straight lines. Knowing that light behaves in this way, and that vision occurs when light enters the eye, Al-Haytham posited an early theory of how the eye itself functions. He based his anatomy of the eye closely on the teachings of the Greco-Roman physician and philosopher Galen, but expanded his predecessor's ideas by observing how different factors, including eye movement and psychological phenomena, can affect perception. Al-Haytham's work went on to influence numerous other scholars, and his commitment to systematic experimentation and the scientific method helped to establish the tenets of science more broadly. Several centuries after his death, Al-Haytham's theories were translated into Latin and began to gain a following in the West. Western scholars adopted and built on his ideas, supporting the growth of the field in Europe, where it would go on to prosper during the Renaissance.

Who's Who – Ibn Al-Haytham

The mathematician and astronomer Ibn al-Haytham was a monumental force for science in the Islamic Golden Age, and optics was just one part of his impressive body of work. Born in Basa, Iraq, around AD *965, Al-Haytham lived during a time in which Islamic science flourished. One account of Al-Haytham's life suggests that he moved to Egypt at the behest of a ruler known as 'the Mad Caliph', who wanted him to prove a claim Al-Haytham had made that he could regulate the flooding of the Nile. When Al-Haytham realized that this was not possible, he began to fear for his life. He therefore feigned madness and remained confined to his house until the death of the caliph in 1021.*

Over the course of his life, Al-Haytham wrote up to 200 books, many of which went on to influence European scientific thought. His commitment to the scientific method – centuries before it became widely popularized – pre-empted the shift towards a more modern way of conducting research. In recent years, Ibn al-Haytham has begun to receive some of the recognition he deserves for his extensive contributions, which undoubtedly earn him a place in the history books.

THE ROAD TO MODERN OPTICS

Optics gained much ground during the centuries after the post-classical period, as new technologies and new ideas opened up the field. At the turn of the 17th century, a German mathematician and astronomer named Johannes Kepler published an essay documenting several optical phenomena in relation to the science of astronomy. Kepler's essay also contained a description of the human eye, which acknowledged the role of the retina in forming images and identified the physical causes of long and short-sightedness.

But the question still remained as to what light actually *is*. In 1690, Christian Huygens, a Dutch physicist, astronomer and mathematician, published his theory of light, proposing that light is a wave that moves in peaks and troughs. After all, scientists had recently discovered the phenomenon of diffraction, which suggests a wave-like nature. But the English polymath Isaac Newton did not agree. In 1704, Newton wrote that light is made up of separate particles that travel in a straight line.

Newton's argument proved most convincing to the scientific community, and the particle theory of light became widely accepted for over a century afterwards. Newton wasn't entirely right when it came to his particle theory of light, although he wasn't entirely wrong either. But Newton's most famous contribution to the field of optics is perhaps his prism experiment, which revealed that white light is made up of many different colours, each with slightly different properties. Newton may not have fully realized it at the time, but his discovery demonstrated that light was infinitely more complex than previously imagined.

Did You Know?

Electromagnetic radiation is a type of energy that can be emitted in many different forms. Visible light is just one of several forms of electromagnetic radiation; radio waves and X-rays are others. The electromagnetic spectrum orders the various forms of electromagnetic radiation based on their wavelength. Visible light is somewhere in the middle of the spectrum. Radio waves have a very long wavelength, so they are on one side, while short-wavelength gamma rays are on the other side.

In the 18th and 19th centuries, Newton's particle theory fell out of favour. Scientists such as Thomas Young conducted experiments which showed that light behaves more like a wave than anything else. Scientists also began to realize that visible light is only a small part of the full spectrum of light. Ultraviolet and infrared light were discovered, each with strange properties unlike the light we're used to. In 1865, a Scottish scientist named James Maxwell formulated a new theory of light, proposing that electricity, magnetism and light are all part of the same phenomenon. Arguing that visible light itself was an electromagnetic wave, he predicted that there were probably other types of similar electromagnetic waves. Maxwell was absolutely correct, and in the years that followed, scientists would realize that visible light is just one of many forms of electromagnetic radiation.

Isaac Newton performed his legendary prism experiment more than 300 years ago.

In the early 20th century, the great German physicist Albert Einstein breathed new life into Newton's particle theory of light, when he observed that light may exhibit both particle and wave-like properties at the same time. Einstein suggested that light is actually made up of individual physical units, which we now call photons. But these photons are very strange, because they actually behave like waves and particles at the same time. This quality is known as wave-particle duality, and it accounts for many of the strange phenomena we see in optics, whereby light doesn't always behave in predictable ways. In recent years, scientists have begun to learn more about the quantum nature of light, and the ways in which it can seemingly shift behaviour and act in two conflicting ways at the same time.

Did You Know?

When light is shone on to a metal, electrons are sometimes emitted from the material's surface. Scientists observed this effect in the 19th century, but it was Albert Einstein who finally worked out that the phenomenon resulted from light particles – or 'photons' – hitting the electrons and dislodging them.

From the use of quartz crystals as reading lenses by ancient civilizations to the building of telescopes capable of exploring the first stars and galaxies of the universe, the exploration of optics has been empowering humankind for centuries. Even now, centuries on from when the first scientists began to question the nature of light, we still remain uncertain. Optics is a large and complex field, and there is still a great deal of knowledge left to uncover.

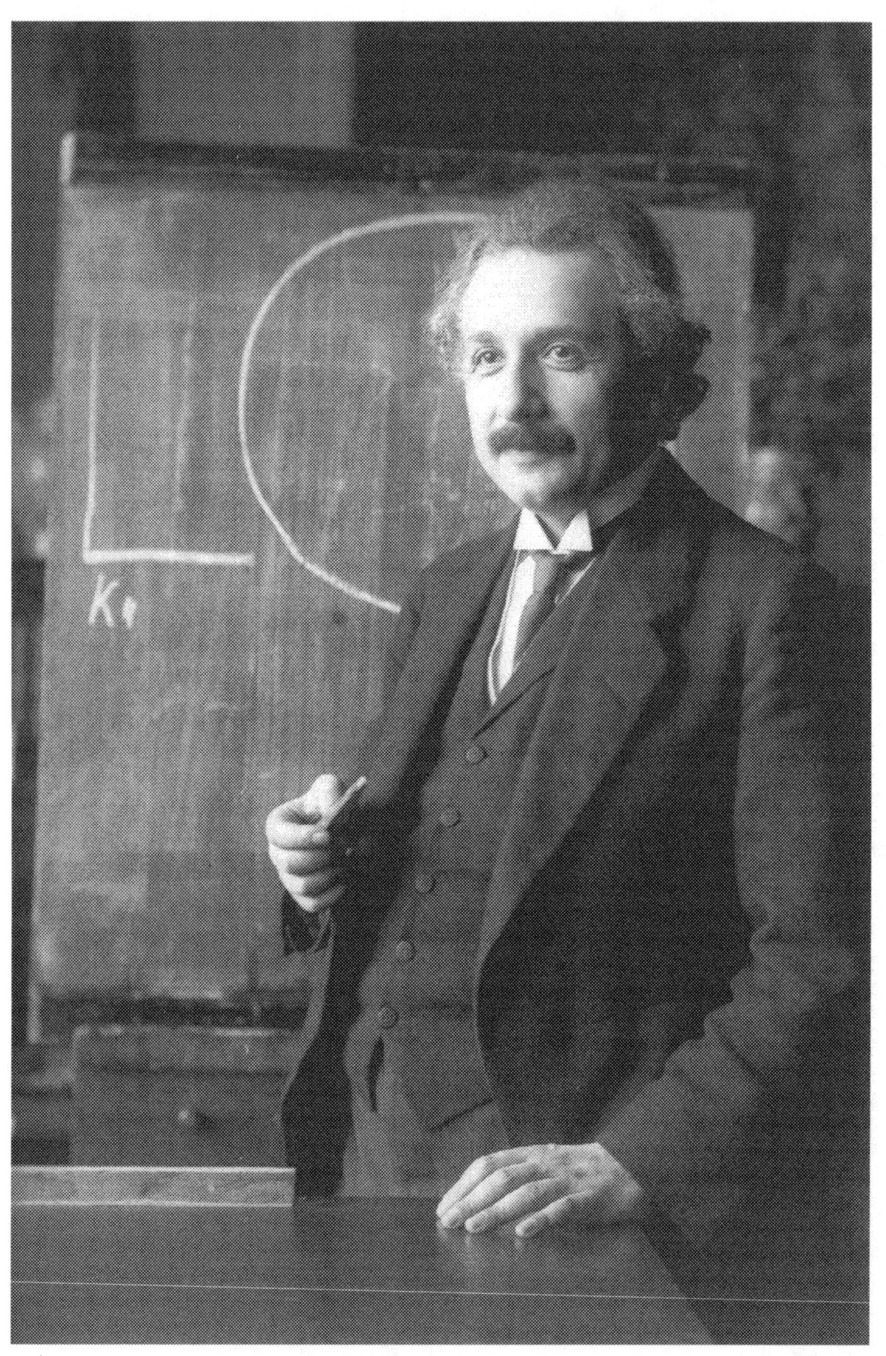

In 1905, Albert Einstein suggested that light behaves both as a wave and as particles, a quality known as wave-particle duality.

CHAPTER 6
PLANT SCIENCE

'There is not a sprig of grass that shoots uninteresting to me.'
– THOMAS JEFFERSON, US PRESIDENT (1801–09),
STATESMAN, DIPLOMAT AND FOUNDING FATHER

The progress of humankind is intimately tied up with the existence of plants. From the food that we consume, to the medicines we take, to the materials we use, plant matter plays a fundamental role in all of our lives. And this has always been the case, right from the beginnings of human society.

For a long time, plant science was dominated by its applications to agricultural and medical science, but over the centuries it has grown into a scientific discipline in its own right. Plants are the planet's dominant species, accounting for 80 per cent of the total biomass on Earth; they provide most of the world's oxygen and are fundamental to the survival of the human race. So plant science is deeply important; the most pressing challenges faced by humankind in the 21st century – food security, global health and biodiversity – all intersect with the science of botany.

Plant science has a long and curious history that stretches across many centuries and civilizations. And, as with so many scientific fields, it's a history that is still being written. So let's dig into the history of botanical thought, and the story of humans' long and complex relationship with plant life.

c. 10,000 BC	*Neolithic tribes transform into agricultural communities*
c. 4th–3rd CENTURY BC	*Theophrastus establishes early principles of botany*

8th CENTURY	*Islamic world begins to undergo an agricultural revolution*
12th CENTURY	*Hildegard von Bingen establishes holistic theory of medical botany*
13th CENTURY	*Ibn al-Baitar publishes influential treatise on medicinal plants*
1440	*Invention of the printing press triggers surge in botanical publishing*
1543	*First botanical garden established at University of Pisa in Italy*
1665	*Robert Hooke publishes microscopic studies of plant tissues*
1753	*Carl Linnaeus develops system of plant taxonomy*
1994	*First genetically modified crop becomes commercially available*

THE ROOTS OF PLANT SCIENCE

Plant science has origins that stretch back into the depths of time. When humans were still hunter-gatherers, botanical knowledge was key to survival. But our relationship with plants grew more complicated with the dawn of agriculture around 10,000 BC, as previously nomadic tribes began to form settled communities. Agriculture first took root in the Fertile Crescent, but it also developed independently in different parts of the world, with separate groups of people turning to farming at various times. While agriculture itself wasn't a scientific pursuit until much later on – people initially turned to it in order to survive – the techniques that the first farmers used to improve their crop yields and food supply did require a rudimentary form of botanical knowledge and observation. This same approach allowed early farmers to breed stronger crops, and to domesticate various plant and animal species.

Did You Know?
The Neolithic Era lasted from around 10,000 BC to approximately 3,000 BC, but the actual timings varied a lot in different parts of the word. The dawn of this era marked a revolution in human life, as hunter-gatherer lifestyles began to give way to settled villages and agriculture.

Domestication: unlike taming, domestication takes place over many generations, and alters the DNA of the plants and animals involved.

Around 3,500–3,200 BC, evidence emerges of a more systematic approach to botany. Early botanical works have also been traced back to Ancient India, Egypt, China and Mesopotamia. However, the foundations of plant science are generally ascribed to Ancient Greece, and the philosopher Theophrastus. A student of Aristotle, Theophrastus was one of the first scholars to study plant science for its own sake, rather than as a tool to support agriculture or medicine. He recognized that plants could be annuals, biennials and perennials, and understood variations in the reproductive organs. He produced detailed notes on the appearance and form of many plants, and carefully studied the anatomy and physiology of plant life. For these reasons, Theophrastus is sometimes referred to as the father of botany. He is estimated to have written around 200 botanical works, but only two survive. Somewhere between 350 and 287 BC, Theophrastus authored *Enquiry into Plants*, which covered structure, reproduction and development, and introduced an early system of classifying and organizing plants based on their characteristics. One of the most influential natural history texts ever written, this text would become a reference point for botanists for centuries to come.

But while plant science was arguably born in Ancient Greece, scientific progress in the field during the centuries to come would originate from the Middle East.

POST-CLASSICAL PLANT SCIENCE

Following a pattern that we've seen in other scientific fields, progress in plant science waned in Europe and prospered in the Islamic world during the post-classical period. Between the 8th and 13th centuries, Islamic scientific understanding of agriculture progressed rapidly as new species and techniques spread along booming trade routes. The Islamic

A page from an ancient manuscript called De Materia Medica *by Dioscorides depicts a physician preparing a medical elixir. It was written between* AD *50 and 70.*

world underwent its own agricultural revolution, as crops and techniques from as far afield as Africa, China and India became diffused throughout Islamic regions, and advances in irrigation and mechanized labour led to increased production, prosperity and population.

Did You Know?
The Fertile Crescent spanned many different countries, touching on portions of what is now Turkey, Syria, Lebanon, Jordan, Palestine, Israel, Iraq, Kuwait, Egypt, Cyprus and Iran.

In the 13th century, a scholar emerged who would become the Islamic world's most renowned botanist. Heralding from Muslim Spain, Ibn al-Baitar is remembered for his vast knowledge of plants from many different areas of the world. He worked as chief herbalist for the Sultan al-Kamil, and travelled through parts of Europe, North Africa and much of the Middle East. As such, his treatise on medicinal plants – *Compendium on Simple Medicaments and Foods* – is over 900 pages long and contains descriptions of 1,400 different species. While much of Al-Baitar's treatise was compiled from the works of earlier scholars, he did make a number of original contributions. The text also touches on the field of chemistry, providing a guide to the production of oils and perfumes using botanical materials. But perhaps the most significant aspect of Al-Baitar's work was his commitment to the scientific method. Empiricism and an experimental approach were at the core of his approach to research, making Al-Baitar's work one of the earliest examples of the scientific method being applied to botany.

During this time, plant science in Europe made slower progress, but it was not an entirely lost period. The 12th-century German abbess Hildegard von Bingen made significant contributions to the field by establishing a holistic theory of medical botany that saw plant life as interconnected with and integral to human life. In the 13th century, a German friar and bishop known as Albert the Great published a seven-volume book on botany. Far from being a dark age for science, the post-classical period did see progress in plant science. Scientific growth and agricultural advancements in the Islamic world, prosperous trade routes and investment in the medical potential of plants all contributed to advances in the field, creating a solid foundation for plant science to grow into its own discipline during the centuries to come.

Who's Who – Hildegard von Bingen

Polymath, visionary, nun: Hildegard von Bingen lived an extraordinary life. Born in Germany in 1098, Von Bingen began experiencing visions from a young age. Although she was still a child, her parents volunteered her to become a nun – a vocation she would uphold for the rest of her life. As with all things she dedicated her mind to, Von Bingen was very successful in the church; she rose through the ranks and was ultimately recognized as a saint by some branches of the Catholic church.

But on top of her religious status, Von Bingen was also an accomplished herbalist and medical authority. During her lifetime, much of the duty of caring for the sick fell to religious groups, and as such, Von Bingen became experienced in the diagnosis and treatment of disease. Her proficiency in medicine was supplemented by her role managing the monastery's herb garden and by prolific reading at the monastery's library. Thanks to this work, Von Bingen gained an impressive knowledge of botany, which informed her work as a healer. She held that all things in nature existed to support humanity, and as such, looked to plants as a primary source of medicine. Von Bingen recorded her scientific knowledge of plants and medicine in books that variously catalogued plants and their therapeutic uses, and laid out her theory of the human body as interconnected with nature.

Von Bingen is recognized today as one of the most accomplished female thinkers of the Middle Ages. Her pairing of botany and medicine encapsulated the importance of plant science in medieval Europe as a tool for healing. Also, having lived during a time when women's scientific contributions were often ignored or actively discouraged, Hildegard von Bingen's determination to make her voice heard is worthy of much admiration.

THE ROAD TO MODERN PLANT SCIENCE

The refinement of the printing press in the 15th century had a marked impact on plant science. Across Europe, botanical texts known as 'herbals' were published in great numbers. These texts, which expounded the medical properties of different plant species, provided observations of new species along with detailed descriptions and illustrations. As time went on, some herbals began to focus more on the plants themselves rather than their medical uses, and by the 17th century, herbals had evolved into 'floras': academic studies of plants, often accompanied by physical samples kept in a herbarium. These texts demonstrated a shift away from the long-established conflation of botany with medicine. Scholars began to study plants for their own sake, just as Theophrastus had done centuries before. Plant science had finally become a discipline in its own right.

During the European Renaissance, interest in botanical study grew, as trade, travel and the popularity of botanical gardens generated curiosity about new plant species. The invention of the compound microscope towards the end of the 16th century heralded a new chapter in the history of plant science. Prior to this, botanists had relied on what they could observe with their eyes; now, they could look deeper into the microscopic structure of plants. In 1665, the English polymath Robert Hooke published *Micrographia*, a textbook containing sketches of objects as seen through the lens of a microscope. Hooke's sketches provided unprecedented insights into the natural world, revealing the imperceptible intricacy behind everyday things such as a fly's wings or a flea. Hooke also included studies of plant tissues in his book. Looking through a microscope, he observed for the first time microscopic holes in slices of cork, which he labelled cells, and observed that living plant cells contain other materials, including sap. Hooke's book laid the foundations for the development of plant anatomy in the 17th century. Plant anatomy involved exploring the internal structure of plants and observing characteristics at the microscopic level. By the mid-18th century, another subdiscipline of plant science had emerged: plant physiology, which focused on the biological processes taking place within plants, such as respiration and photosynthesis.

Did You Know?

Botanical gardens are spaces which house and cultivate a broad range of different plant varieties as part of a 'living collection'. Beginning as an aid to medical research at universities, botanical gardens have a long history in the science community, although many botanical gardens are now freely open for the wider public to explore.

As botanists began to learn more about plants and uncover more and more species around the world, the system for organizing and classifying these plants became increasingly cumbersome and convoluted. Different botanists used different systems, so one species could be known by a variety of different names, which made it difficult for plant scientists to collaborate and share information. All this changed in 1753, when a Swedish naturalist named Carl Linnaeus published a treatise establishing the modern method of plant taxonomy. Linnaeus' system used two Latin words: the genus name and a specific name to identify a plant. For instance, English lavender would be known as *Lavandula angustifolia*. This created a clear system that botanists could use to classify and study plants.

Progress in plant science continued in the 19th century, as research began to intersect with other areas of scientific study, including Charles Darwin's studies of evolution, Gregor Mendel's investigations into heredity, and Alexander von Humboldt's work on geography (see page 62). By the close of the century, plant science had grown into a large and

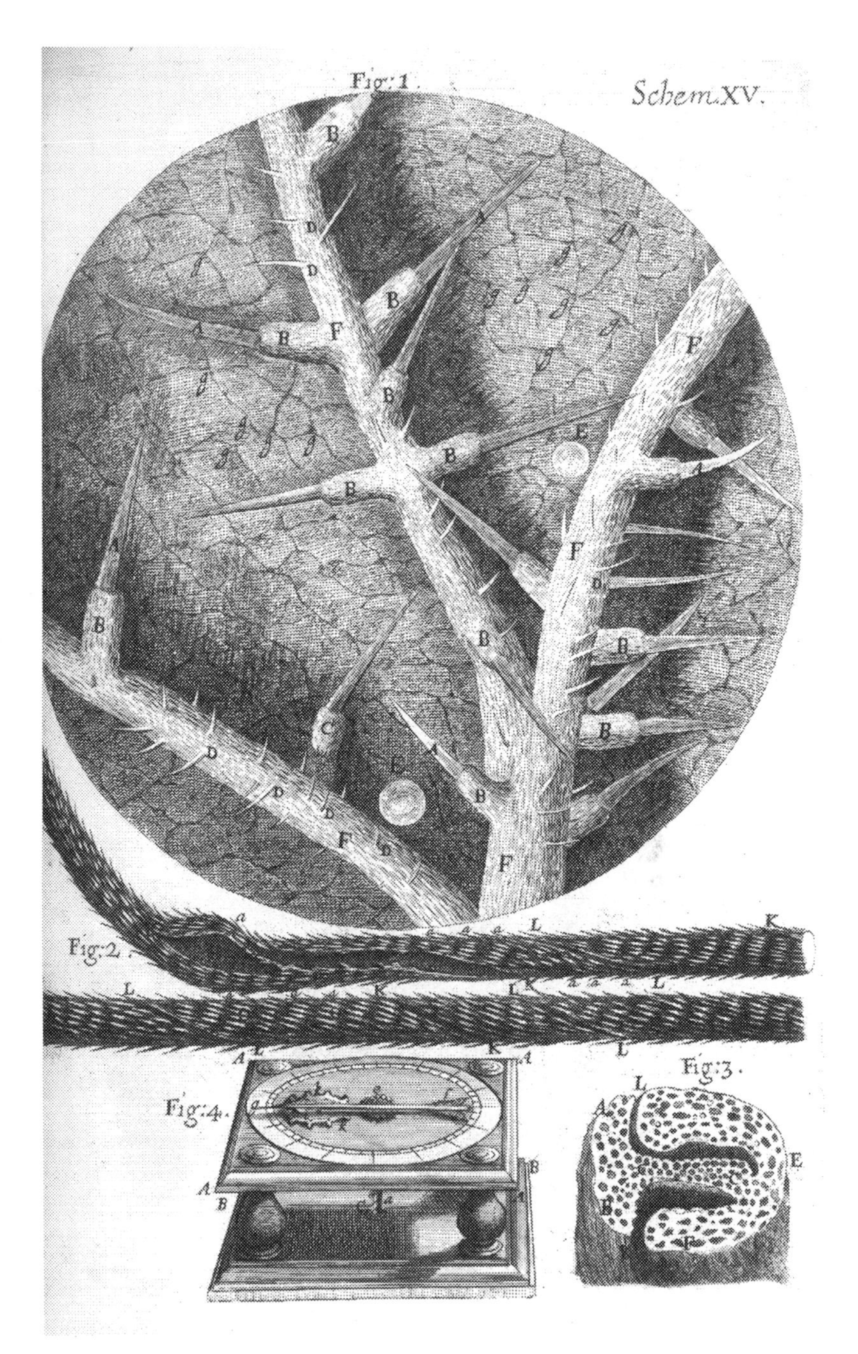

An illustration from Robert Hooke's Micrographia. *A study of minute objects seen through a microscope, the book, which was first published in 1665, provided unprecedented insights into the natural world.*

multi-faceted discipline, touching on many different areas of science.

We have come a long way in the study of plant science, but like our ancestors before us, our existence is still very much intertwined with that of plant life. In the 20th and 21st centuries, botanical research has increased in both volume and scope, as scientists continue to explore the complexities of plants and their applications to human life. Plants remain fundamental to human health, having supported the development of antibiotics and other vital medicines, and scientists today continue to use plant matter to support experiments in laboratory settings. We are

The geneticist Gregor Mendel's investigations into heredity played a major role in plant science.

also still very reliant on plants for sustenance, and new insights into the genetics of plants could help to improve access to nutrition in regions where food sources are scarce or unreliable. What's more, in the future the knowledge derived from plant science will be fundamental to helping maintain global food security and fragile ecosystems as the climate changes around us.

Ultimately, plant science is still just as relevant to us now as it was to our Neolithic ancestors all those years ago. Like them, we still have a long way to go before we know everything there is to know about plant life on Earth. Plant science is no longer merely the observation of species or the collation of herbal remedies; the discipline has grown into a huge field with applications in many crucial areas of human life. Our fate as a species is intimately connected with that of plants, and understanding plant life and its applications is perhaps more important now than ever before.

Plant Breeding

When you take a bite into a big, crunchy corn on the cob, what you're tasting is the product of centuries of breeding and development. For thousands of years, people have been carefully selecting the most desirable characteristics – size, taste and convenience – to create the version of corn that we now see on our dinner plates. That progress began with our ancient ancestors observing and experimenting with their harvests.

Prior to the Neolithic Revolution, nomads gathered and ate wild grains that they found growing on trees and bushes, but when humans became settled, they had time to begin cultivating crops such as wheat and barley. The first farmers started off by planting and harvesting the same seeds that grew naturally in the wild, but after a while they discovered that they could get more food if they were smart about which crops they chose to grow.

These days, plant breeding is much more advanced, but in the Neolithic Era, people didn't know anything about genetics – it would be millennia before the likes of Gregor Mendel and later

Charles Darwin began experimenting with plant inheritance. But through empirical observation, farmers figured out that when they planted seeds produced by the highest-performing crops, the quality of the harvest would gradually improve. And so, each year, early farmers would plant seeds from the biggest, healthiest and most reliable crops, and in turn, their plants became bigger, healthier and more reliable, until eventually they became fully domesticated.

Today, plant breeding is more likely to involve either artificially controlling the mating process between different strains or else directly manipulating the genetics of a strain to promote the desired characteristics. But we wouldn't have these shortcuts had it not been for those observant Neolithic farmers of the past. Neolithic humans didn't set out to become plant breeders, but their innate curiosity, their willingness to experiment and their drive to survive in a hostile environment sparked a process that remains fundamental to agriculture today.

PART THREE

THE EARLY MODERN PERIOD

15TH–18TH CENTURIES

THE EARLY MODERN PERIOD was a time of rapid advancement and change. Over the course of roughly 400 years, Europe underwent the Renaissance, the Reformation, the Scientific Revolution and the Enlightenment that followed.

A key feature of this era was its globalizing character – exploration became a political imperative for nations in Europe at this time, leading to an increase in sea voyages and cross-cultural contact. European colonization of 'newly discovered' lands in the Americas and the establishment of outposts in Southeast Asia and South Africa supported the growth of new trade routes. Furthermore, the exchange of foods, products and slaves between the Old World of Europe and the New World of the Americas saw some Western European nations become increasingly powerful and wealthy.

The early modern period was also a time of growth for the science community. Works by the classical thinkers of Ancient Greece and Rome passed through the Islamic world into Europe, leading to a resurgence in scholarship. At the same time, attitudes towards science began to change, and the pursuit of knowledge gradually came to be regarded as a more valuable endeavour than it had been in the past, resulting in more patronage for scientific research.

Thanks to this increased focus on science and research, the formal scientific method became established and widely adopted during the early modern period, which in turn led to a surge in scientific progress. Meanwhile, the authority of the church gradually began to decline, as trust in science and evidence continued to grow. We see the impact of these wider shifts manifested in the scientific revolution that took place between the 16th and 17th centuries. During this period, thinkers developed new theories about nature and the natural laws of the universe, ushering in a new, altogether more modern phase in human history.

As a result of all these factors and more, curiosity and exploration were able to flourish during the early modern period. And, of course, wherever curiosity is nurtured, knowledge and understanding are bound to follow.

CHAPTER 7
ANATOMY

'Anatomy is the great ocean of intelligence upon which the true physician must sail.'

– JOHN E. LINK, SURGEON, QUOTED IN THE JOURNAL OF THE AMERICAN MEDICAL ASSOCIATION IN 1893

Anatomists study the structure and relationship between body parts, work that has historically been fraught with superstition and taboo. Long before the advent of X-rays and microscopes, before cultural practices made it acceptable for scientists to carry out dissections and anatomical experiments, our own bodies were a mystery to us. For centuries, restrictive attitudes to anatomy bred ignorance about the structure and physiology of human beings, leading to the establishment of false dogmas. But the spirit of human curiosity eventually won out, leading to a revolution in anatomical understanding in the early modern period. In the years that followed, a deeper understanding of anatomy helped scientists to uncover new information about disease, to cultivate more effective medicine and to explore the complex processes and structures that underpin human life.

c. 4th–3rd CENTURY BC	*Herophilus and Erasistratus pioneer field of anatomy*
c. 2nd CENTURY BC	*Galen publishes anatomical observations based on animal dissections*
1242	*Ibn al-Nafis describes pulmonary circulation of the blood*
1315	*Mondino de Luzzi writes world's first dedicated anatomy manual*

1543	*Andreas Vesalius produces anatomy book complete with detailed illustrations*
1628	*William Harvey publishes description of heart and circulatory systems*
1816	*René Laennec invents the stethoscope*
1829	*Invention of achromatic lenses ultimately supports microscopic anatomy*
1839	*Theodor Schwann and Matthias Jakob Schleiden propose cell theory*
1972	*Invention of CAT scan allows study of organs in living patients*

THE ROOTS OF ANATOMY

The earliest known evidence of anatomical study dates back to Ancient Egypt, a civilization known for its advanced medical knowledge. The Edwin Smith Papyrus (see page 26), which was written around 1600 BC, contains information on the heart and blood vessels, along with a variety of other internal organs. However, the Ancient Egyptians did not pursue the systematic study of anatomy; their limited knowledge was gained with the aim of supporting medical diagnosis and treatment.

Centuries later, Egypt's capital, Alexandria, would become a centre for learning in the dual disciplines of anatomy and physiology. During this time, the country was ruled by Greece. Dissection had long been forbidden in the Greek Empire, but for a short period between the 4th and 3rd centuries BC, the ban was lifted. It was during this window that two physicians of Alexandria, Herophilus and Erasistratus, pioneered the field of anatomy. In studying human anatomy by human dissection, the pair were able to rapidly progress understanding beyond what had previously been gleaned purely through animal dissection and conjecture. Between them, the two scholars identified the brain as the centre of the nervous system, investigated the structure and function of the heart, and shed light on several intricate biological processes. Unfortunately, despite this brief moment of enlightenment, taboos around human dissection persisted, and progress in the following centuries was stymied.

Physiology is concerned with the way in which the bodies of living organisms work.

The next major advances came thanks to the physician and philosopher Galen. Practising in Rome in the 2nd century BC, Galen was forbidden to dissect human beings, so much of his research was based on animal dissections, including pigs and monkeys. Even with these limited resources, Galen made remarkable progress. He uncovered new information about the kidney and bladder, as well as the respiratory system. In addition, through his experiments on pigs, Galen learned of the significance of the brain in controlling muscles and movement, establishing an early understanding of the nervous system. He also worked out how arteries functioned and identified the differences between them and veins.

However, Galen's reliance on animal specimens meant that he made a lot of errors in his theories about the human body, incorrectly assuming several equivalences between human and animal biology.

In the centuries following his death, Galen became a revered figure and his influence pervaded throughout the medieval Arabic world and Europe, but veneration for his scholarship often came at the expense of critical analysis. Much of what Galen theorized was taken at its word, without acknowledgment of the flaws in his methods and conclusions. In fact, it took over a thousand years for some of Galen's erroneous theories to be challenged. Without challenge, knowledge was unable to grow, and the development of anatomical thought largely stalled. It would take the work of medieval Islamic scholars and the European Renaissance that followed to see anatomy flourish once again.

EARLY MODERN ANATOMY

Anatomy suffered through a period of stagnation for many years after antiquity. Rather than performing human dissections first-hand, scholars looked to the past for their anatomical knowledge, allowing flawed theories to persist for many centuries.

That said, there was a certain amount of progress. In the 13th century, the Syrian-Arab physician Ibn al-Nafis successfully described the

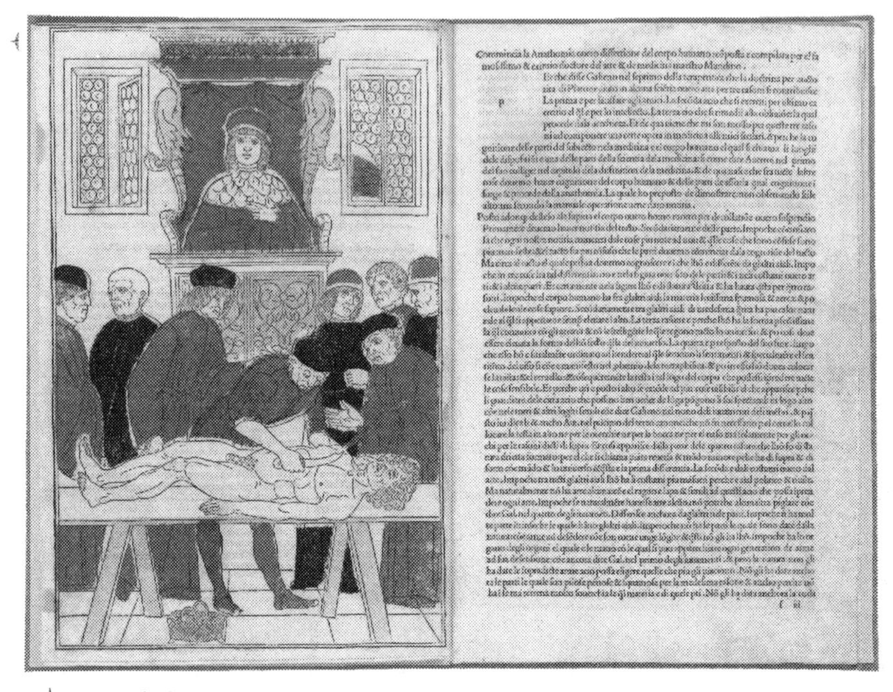

Fasciculo di medicina, *anatomy lecture, 1493. Claudius Galen (AD 129–199) was the most prominent physician in Ancient Greece. His conclusions were purely based on the study of animals, and his faulty theories on human anatomy dominated and influenced medical science until the Renaissance, over a thousand years later.*

pulmonary circulation of the blood, challenging several key assertions made by Galen. Al-Nafis found that blood moves from the right to the left side of the heart through the lungs, not by passing through invisible pores as Galen claimed. He also theorized the existence of capillaries and identified the expansion and contraction of arteries. Al-Nafis' genius went largely unrecognised in Europe during his lifetime and for hundreds of years afterwards. It was only in the early 20th century that his work was rediscovered and the extent of his contributions to anatomy realized.

Who's Who – Ibn Al-Nafis

Ibn al-Nafis was born in 1213 in Syria, and lived during the latter years of the Islamic Golden Age. He began studying medicine in Damascus aged 16, before moving to Egypt where he lived most of his life. Al-Nafis was a polymath and a product of the advanced age in which he lived. He contributed to a range of fields, including philosophy, theology and law, as well as psychology, medicine and ophthalmology.

As a young man, Al-Nafis began writing his most extensive text, Al-Shamil fi al-Tibb (The Comprehensive Book on Medicine). This encyclopaedia of medical knowledge comprised 80 volumes on Islamic medicine. Al-Nafis intended his encyclopaedia to reach 300 volumes, but he died before it could be completed. Nonetheless, the work is among the most comprehensive medical texts ever authored by a single scholar.

At the age of 29, Al-Nafis published his theory of pulmonary circulation. In doing so, he was challenging the authority of both Galen and the respected Islamic scholar Ibn Sina (see page 32). Al-Nafis' willingness to question the accepted wisdom of the times helped advance Islamic medical scholarship well beyond the medieval European tradition.

There is debate as to whether Al-Nafis' work was available outside the Islamic world and whether his writing might have influenced thinkers like William Harvey, who published his own theory of pulmonary circulation several hundred years later. What is certain, however, is that for centuries, Al-Nafis' contributions to science went largely unrecognised in the West. In the early 20th century, his writings were rediscovered, and Al-Nafis is now rightly recognized as one of the greatest physiologists of his age, and a giant in the field of anatomy.

Despite his formidable knowledge of the heart, Al-Nafis claimed to have never performed a human dissection, a practice that was forbidden by Islam during his lifetime, as it was in much of the world. Because of these widespread prohibitions, systematic dissection was rare during the centuries between antiquity and the Renaissance, but it wasn't completely unheard of. In the early 14th century, an Italian physician and professor named Mondino de Luzzi published the world's first dedicated anatomy manual, based on first-hand experience in dissection. De Luzzi's publication was significant because although it repeated many of the erroneous assumptions made by Galen several centuries prior, the book reintroduced the concept of practising systematic dissection to aid anatomical study.

But the real turning point for Renaissance anatomy came in 1543, with Andreas Vesalius' publication *On the Fabric of the Human Body.* A Flemish physician and professor, Vesalius is generally regarded as the father of modern anatomy. His seminal text contained exquisite illustrations of skinless human bodies posing in rural landscapes, their bare muscles and insides exposed in all their intricacy. Vesalius performed human dissections himself – a rarity for anatomists at the time – and his extensive first-hand experience helped him to identify some of the errors in Galen's theories. He proved that the human jaw was made up of a single bone rather than two as Galen had claimed, and he demonstrated that humans do not have the same network of blood vessels at the base of the brain as pigs, sheep and cattle.

Galen's authority was widely accepted by most scholars, and so Vesalius was the subject of much persecution for his work. But he encouraged his students to carry out their own dissections and test his theories directly, rather than accepting either his or Galen's teachings. As a result, more and more people were able to verify the truth of Vesalius' claims. This was a seismic shift; anatomy had passed into the hands of a new generation of scientists with the tools to establish the truth for themselves.

The Scientific Revolution

During the early modern period – roughly spanning the late 15th century to the late 18th century – scientific thought underwent a transformation in Europe, leading to the dawn of a new era. By the end of the Scientific Revolution, modern science as we know it had become established as a set of disciplines. During this time, science came to replace religion as the recognized basis for knowledge about

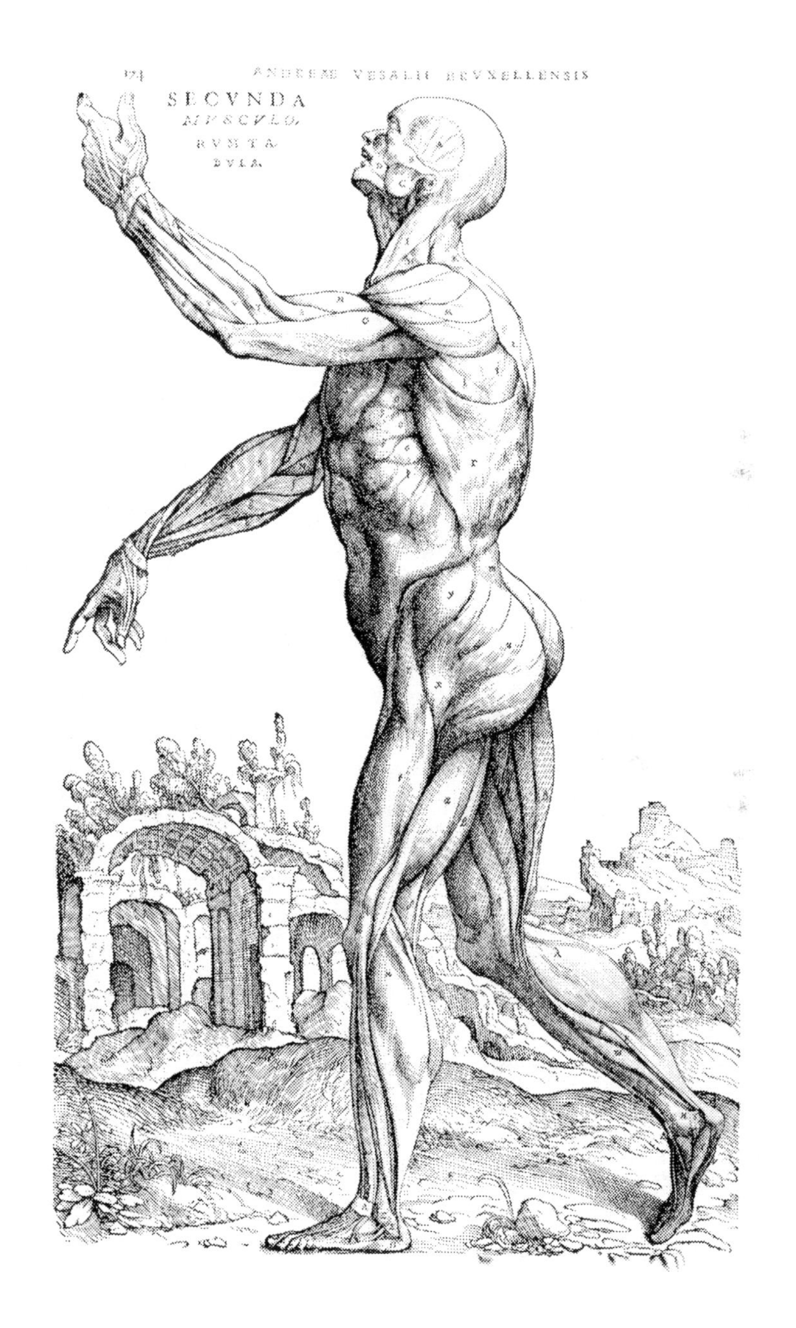

Vesalius revolutionized anatomy with his artistic vision.

the world; meanwhile, the scientific method – with its commitment to experimentation and evidence – helped to establish the sciences' independence from philosophy.

While the exact dates of the Scientific Revolution are a topic of debate, many scholars recognize Nicolaus Copernicus' publication of De Revolutionibus Orbium Coelestium (On the Revolutions of the Heavenly Spheres) in 1543 as a starting point. The work challenged the presiding belief system about the nature of the universe, and unsettled the hegemony of the Church. What followed was a transformation in thought and ideas about the structure and function of nature. The new scientific view did not place humankind at the centre of a divine universe, but recognized the existence of fundamental laws of nature that could be quantifiably measured and understood.

The Scientific Revolution was not only groundbreaking for the sciences; it also created a seismic shift in wider intellectual thought. Philosophy, politics and the arts all underwent a transformation in the Age of Enlightenment that followed. The scientific commitment to reason reverberated on the political stage, as ideas such as liberty, freedom and the separation of church and state began to gain pace, ultimately culminating in revolutions in France and the United States.

Scholars often place the culmination of the Scientific Revolution in 1687, with Isaac Newton's publication of Principia *– a text which laid the foundations for classical physics and informed scientific thought for almost 300 years. As the revolution came to an end, scientists were left with a new, deeper level of understanding about nature, and a clear process to help generate reliable theories that expanded that knowledge. A new era of experimentation and reason had taken hold, and the foundations had been set for the growth of modern science as we know it.*

The compound microscope improves on simple microscopes – like magnifying or reading glasses – by using two types of lens to create a more highly magnified image.

Barely half a century after Andreas Vesalius' death, the English physician William Harvey published a comprehensive description of the workings of the heart and circulatory systems – work that is all the more impressive for the fact that it was done without the use of a microscope. Indeed, it was the introduction of the compound microscope in the 17th century that shifted the course of anatomical history by allowing scientists to study biological structures that couldn't be seen

The English physician William Harvey was the first to explain blood circulation in the human body. Here, he is depicted explaining the workings of a deer heart to Charles I.

with the naked eye. Researchers such as the Dutch natural philosopher Jan Swammerdam and the Italian physician Marcello Malpighi helped to expand the field of microscopic anatomy, and by the end of the century, the discovery of capillaries, blood cells and other previously invisible biological matter had heralded in a new age in anatomical study. For the first time, anatomists became aware of an even deeper layer of complexity underpinning the structure of human life.

THE ROAD TO MODERN ANATOMY

In the years that followed, the study of anatomy became more systematic and grew in depth, as increasingly powerful technology helped scientists to dig ever deeper into the intricacies of the body. As anatomical knowledge grew, a more modern approach to medicine began to emerge, based on a detailed understanding of human physiology. In the early 1800s, the inventor the stethoscope, French physician René Laennec helped to bring the fields of anatomy and pathology closer together. Using his device, Laennec was able to relate the sounds he heard when listening to his patients' chests to physical irregularities identified inside their bodies during autopsy. In this way, Laennec connected anatomical abnormalities with their corresponding disease, helping to advance new anatomical and physiological approaches to diagnosis.

Did You Know?

René Laennec reportedly created the stethoscope in part to avoid the social awkwardness of having to put his ear next to female patients' chests.

In the 1830s, achromatic lenses were introduced, dramatically enhancing the performance of microscopes. With this improved technology, a pair of German researchers named Theodor Schwann and Matthias Jakob Schleiden were able to explore biological cells. Schwann and Schleiden proposed that living organisms are composed of cells, which are the basic units of life, and that these are generated from pre-existing cells through cell division. This discovery solidified a new era in anatomy; one marked by the study of the very small structures of life. As anatomy shifted further to the microscopic scale, new fields branched off, such as cytology, which is the science of cell biology, and histology, which studies the microscopic structure of biological tissues.

Achromatic lenses are specially shaped to manipulate the wavelengths of light in a way that reduces distortion and aberration of the image.

The advance of anatomy has remained closely tied to technological developments. The advent of X-ray technology in the late 19th and early 20th centuries helped scientists to learn more about the skeletal structure of the body. In particular, the development of the electron microscope – which uses electrons rather than visible light to illuminate a sample – helped to expand the field dramatically. From the 1950s onwards, this device allowed anatomists to study biological structures at the subcellular level. Similarly, the invention of medical MRI machines and CAT scans in the 1970s allowed researchers to explore the anatomy of organs in living patients. Thanks in part to the work of these anatomists, we now recognize the profound significance of the structure and function of molecules like DNA to living things. Much of modern anatomy now focuses on the microscopic and submicroscopic scale, and anatomists continue to expand our understanding of the human body by looking at the very small building blocks of life.

Did You Know?

In 1858, an English anatomist named Henry Gray published an anatomy manual for students. His book would become the most widely known text on the subject of anatomy, and is still in use today. Unfortunately, the eponymous author of Gray's Anatomy *would not live to see the success of his book. He died at the age of 34, having contracted smallpox after treating his sick nephew.*

The French physician René Laennec.

Anatomy has had a long and interesting journey through history, and the story of anatomy's suppression and eventual rebirth is testament to the importance of active experimentation and interrogation as cornerstones of scientific enquiry. As Vesalius taught us, it's only by directly testing theories that we can be sure of what is and what is not true. It is through this systematic approach to anatomical study that scientists through the ages have helped us to improve diagnosis and medicine, and to better understand the exquisite physical structures that allow living organisms to function.

CHAPTER 8
ASTRONOMY

'Astronomy is the most ancient of all the sciences, and has been the introducer of vast knowledge.'

– MARTIN LUTHER (1483–1546), THEOLOGIAN,
COMPOSER, PRIEST, MONK AND LEADING FIGURE
IN THE PROTESTANT REFORMATION

Since the dawn of human life, people have looked to the skies for answers about the nature of existence and our place in the world. Astronomy is perhaps the oldest form of science, and has featured in human societies since prehistoric times. We now understand astronomy to mean the scientific study of the universe and of celestial objects such as planets and stars, but for many years it was a tool for predicting the future and worshipping the gods. This gave the discipline a prized place in many ancient cultures, leading to the growth of astronomical knowledge from the earliest days of civilization. But the ties between astronomy and religion have also led to controversy and the suppression of scientific thought. Over time, new astronomical ideas have caused fractures with the church, challenging the notion of humanity at the centre of a perfectly ordered universe, created by the hands of God.

c. 1200 BC	*Babylonians create early catalogues of the stars*
c. 3rd CENTURY BC	*Aristarchus of Samos proposes first known heliocentric theory*
c. 150	*Claudius Ptolemy argues the Earth is the centre of the universe*
1543	*Nicolaus Copernicus triggers off Copernican Revolution*

1616	*Church bans Galileo Galilei's work*
1609	*Johannes Kepler publishes laws of planetary motion*
1687	*Isaac Newton establishes principle of gravity to explain planetary orbits*
1800	*William Herschel discovers infrared radiation*
1960s	*Astrophysicists develop Big Bang theory*
1969	*Human beings set foot on the Moon*

Did You Know?
The Ancient Egyptians' reverence for astronomical ideas was reflected in the building of the pyramids, which are aligned with the North Star.

It may be tempting to think that astronomy, with its scientific emphasis on the particulars of a vast and uncaring universe, strips some of the wonder and magic from the skies; but in some ways, the very opposite is true. Through the systematic study of astronomy, we have learned that the universe is more vast, more complex and more amazing than our ancestors could ever have imagined.

Did You Know?
The oldest known lunar calendar, discovered in Aberdeenshire in Scotland, is thought to date back to 8000 BC.

THE ROOTS OF ASTRONOMY

Human beings have been methodically observing the cosmos for millennia. Some of the earliest examples of what we would now recognize as astronomy date back to ancient Mesopotamia. From at least 1200 BC, the Babylonians created catalogues of the stars; they recognized that astronomical phenomena occur periodically, and could therefore be mathematically predicted. The Babylonians recorded their observations on a series of 70 tablets, known collectively as *Enuma Anu Enlil*, which document various celestial phenomena, as well as omens and mystical predictions. This work went on to influence future civilizations, including the Ancient Greeks. Greek scholars applied the principles of geometry to astronomical studies, and took on a more scientific approach to studying the stars.

In the 3rd century BC, the Ancient Greek scientist Eratosthenes used mathematics to made a strikingly accurate estimate of the circumference of the Earth, the oldest known measurement of its kind (see page 51). In the same century, the astronomer mathematician Aristarchus of Samos proposed the first known heliocentric theory when he suggested that the

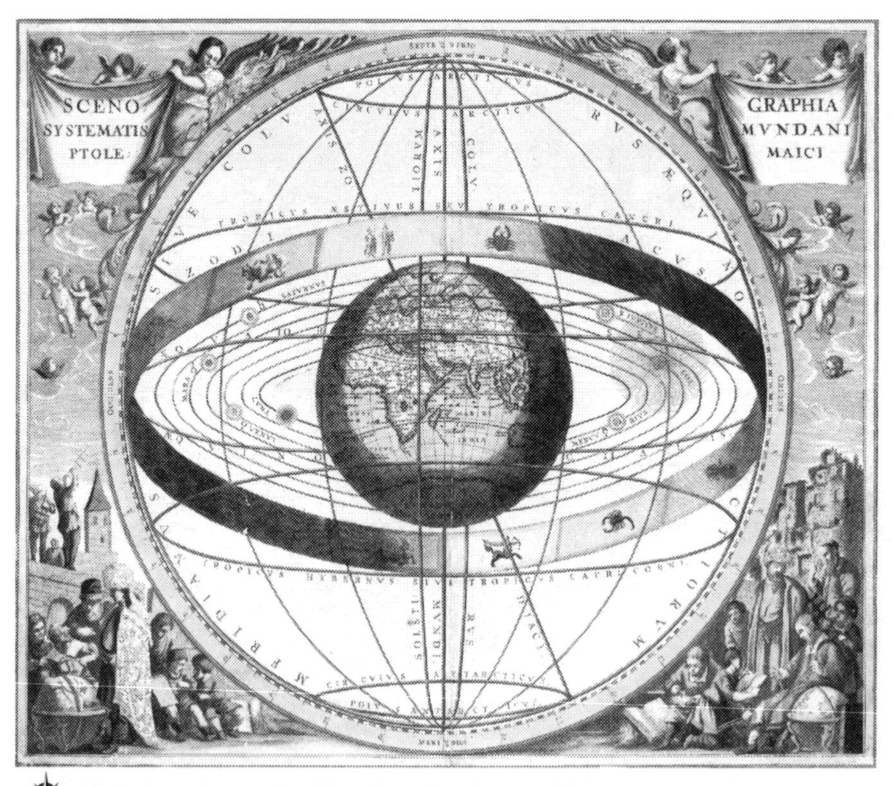

The Ptolemaic (geocentric) model put the Earth at the centre of the universe.

Earth and other planets revolved around the Sun, which sat at the centre of the universe. He also theorized the size of the universe, and proposed that the Earth spins on its axis, accounting for night and day. By placing the Sun rather than the Earth at the centre of the universe, Aristarchus' theory set off a debate that would continue for centuries to come. In Roman Egypt, around the year AD 150, the Alexandrian astronomer mathematician Claudius Ptolemy published his treatise *Almagest* which built on Aristarchus' ideas, but challenged the heliocentric model, arguing instead that the Earth was at the centre of the universe. It was Ptolemy's theory that stuck, and the geocentric model would persist for another 1,200 years.

Did You Know?

In China, detailed astronomical records were kept from around the 6th century BC, and the Chinese were the first to record the occurrence of a super-bright, extremely powerful explosion of a star, now known as a supernova, in AD 185.

During the post-classical period, the Islamic world played an important role in the development of astronomy. Islamic scholars preserved and built upon the work of Ptolemy and other Greco-Roman and Indian thinkers, but they also generated a wealth of new knowledge. Advocates of 'observational astronomy', Islamic scholars championed the importance of direct observation of the skies, and by the 9th century, the Islamic world was home to some of the earliest known astronomical observatories. Although this helped scholars to build their knowledge of the skies, the geocentric model still mostly pervaded. While some Islamic thinkers disputed the idea that the Earth was the centre of the universe, it was not until the European Renaissance that the heliocentric model reached maturity.

The Renaissance is the period of European history that followed on from the Middle Ages. The period was marked by a 'rebirth' of interest in classical scholarship, spanning the arts and sciences.

EARLY MODERN ASTRONOMY

Although astronomical progress had been slow in medieval Europe, all of that was to change during the Renaissance. One of the biggest names associated with this period is the Polish astronomer Nicolaus Copernicus, whose heliocentric theory of the universe finally brought down the idea that the Earth was the centre of all things. Published in 1543, Copernicus' astronomical treatise *De Revolutionibus Orbium*

The Crab Nebula is a supernova remnant of a stellar explosion that took place 6,523 light years from Earth, and was observed by Chinese astronomers in 1054.

Coelestium (*On the Revolution of the Heavenly Spheres*) set in motion a series of events that would ultimately lead to a reckoning between science and religion. Copernicus proposed that the Earth rotates on its axis daily and that it – and all the other planets in the solar system – orbit around the Sun; this put the Sun rather than the Earth at the centre of the universe. Copernicus died from a stroke the very same year that his seminal text was published, so he never witnessed its profound impact. But although Copernicus was not there to see it, his work triggered the dawn of a revolution in thought that bore his name. The 'Copernican Revolution' marked a gradual shift away from the idea of a stationary Earth at the centre of the solar system, towards the model of the cosmos that we now know to be true.

Observational astronomy is the direct study of the skies using telescopes or other optical instruments; while theoretical astronomy develops models to explain the phenomena observed.

Copernicus' heliocentric model posed a challenge to the Catholic Church, who perceived the universe as God's own creation, with humanity at the centre. Although it was powerful and influential enough to cause problems for Copernicus, the Church did not initially react to the new theory. It was only in the early 16th century, when an Italian scientist named Galileo Galilei began popularizing Copernicus' ideas, that the Church began to sit up and take notice. Galileo was an accomplished astronomer who had published observations made through a telescope that he designed himself. Using this device, Galileo was able to discover that the surface of the Moon was rough and cratered. His telescope was so powerful that Galileo was able to estimate the height of mountains on the Moon by studying shadows on the surface. He was the first to observe that the cloudy-looking Milky Way is in fact made of stars, and that the planet Jupiter has several moons orbiting around it. Soon after, Galileo noted that Venus goes through a full set of phases, like Earth's own moon – a phenomenon that would not be visible if Venus were orbiting the Earth. Galileo's discoveries added weight to the Copernican theory of the universe, and as astronomers gradually began to join Galileo and convert to the heliocentric model of the cosmos, the shift in thought was not welcomed by the Church.

In 1616, the Church banned Galileo's work, but the astronomer would not be silenced; he continued to argue that the Earth orbited the Sun. As a result, Galileo was ultimately put on trial and forced to take back his claims; he was sentenced to house arrest, and remained confined to his home for virtually the rest of his life. But his ideas continued to spread across Europe, and by the late 17th century, heliocentrism had become widely accepted.

Who's Who – Johannes Kepler

Johannes Kepler was born in Germany in 1571, at a time when it was still widely believed that the planets and the Sun orbited the Earth. But at university, Kepler studied the work of Copernicus, and found his heliocentric theory convincing. As Kepler's career progressed, he came to work with a renowned Danish astronomer named Tycho Brahe. After Brahe's death in 1601, Kepler studied his data and discovered that Mars orbited in an elliptical fashion.

Between 1600 and 1609, Kepler published his three laws of planetary motion in the solar system. These laws outlined the way in which planets orbit the Sun, demonstrating that orbits are more elliptical than circular, that the orbits of planets speed up when closer to the Sun, and that the further away a planet is from the Sun, the slower it moves. Kepler's descriptions were revolutionary in scope and have since become fundamental to astronomy, but it would be another century before the physics underpinning these laws would be fully explained.

Towards the end of the century, the English physicist Isaac Newton adopted Kepler's laws of planetary motion to support his theory of universal gravitation, arguing that the reason planets' orbits speed up when closer to the Sun is because of the pull of gravity. Though this theory, Newton brought together the two worlds of physics and astronomy, explaining for the first time the physics behind the celestial phenomena.

Galileo used his new device, the telescope, to observe the rings of Saturn.

Towards the end of the 17th century, the work of English physicist Isaac Newton brought together the worlds of physics and astronomy: a feat that is generally viewed as the culmination of the Copernican Revolution. Some years earlier, the German astronomer Johannes Kepler had published three laws of planetary motion which outlined the way in which planets orbit the Sun. Newton's contribution was to explain for the first time the physics behind the orbits of the planets, tying together the observed phenomena and theory with his own theory of gravity. The years that followed were marked by an emphasis on the importance of observational astronomy. Astronomers such as England's James Bradley spent much of their lives on tracking the skies, leading to the discovery of new nebulae, comets and planets.

The early modern period was a climactic time for astronomy; over the course of a few hundred years, astronomers had challenged the authority of religion in matters of the cosmos and established their field as an independent science, grounded in fact and unchained from religious thought.

THE ROAD TO MODERN ASTRONOMY

In the 19th century, methods of observing the skies became more advanced, with an emphasis on precision and careful calculation. At the same time, the links between astronomy and physics grew stronger with the emergence of astrophysics, which began in earnest with the expansion in our understanding of light. In 1800, the British-German astronomer William Herschel discovered that different colours of visible light have different temperatures; at the same time he also discovered that the hottest form of light is not visible at all; we now know that this invisible light is infrared. This was the first in a series of discoveries that revealed visible light to be part of a much larger spectrum of electromagnetic radiation. For astronomers, the discovery of the electromagnetic spectrum would prove revolutionary. Prior to this, scientists could only study celestial objects by using their eyes or a regular telescope, but with an understanding of the electromagnetic spectrum they could study the stars, planets and galaxies on the basis of the other electromagnetic radiation they emit.

Astronomical Spectroscopy

In 1859, a pair of German scientists – one a physicist, the other a chemist – made a breakthrough discovery. Gustav Kirchhoff and Robert Bunsen found that when different elements are heated over a flame, they produce a unique spectrum of light. This spectrum is like a 'fingerprint', an exclusive marker that exists only for that element. This was big news for astronomy. Thanks to this technique, astronomers became able to determine the composition of elements within a celestial object, such as the Sun or the stars, by examining the spectrum of light that the object emitted.

William Herschel's work demonstrated that there are invisible forms of light.

Throughout the 19th and 20th centuries, astronomers continued to develop ways of using the electromagnetic spectrum emitted by celestial objects to study their characteristics. These techniques, known under the collective banner of 'astronomical spectroscopy', have allowed astronomers to study the chemistry, temperature, position, mass and motion of stars, planets, galaxies and more, leading to dramatic gains in our knowledge of the universe.

In the 20th century, Einstein's theory of relativity led to a renewed understanding of both gravity and spacetime. Einstein's elegant theory supported the development of modern astrophysical theories, including the existence of dark matter and black holes. By the 1960s, theorists had used the wealth of techniques and knowledge that astronomers had developed over centuries of research to develop a convincing theory on the origins of the universe, known as the Big Bang theory.

Astrophysics is a branch of astronomy that draws on the fields of chemistry and physics to study the nature of the cosmos.

Midway through the 20th century, we entered into a new era in astronomy as, for the first time, astronauts travelled into space. In the years since, robotic technology has developed to allow us to explore far-flung reaches of the solar system, including comets, moons and planets. Remote telescopes, situated either on Earth or in space, have provided astronomers with insights into previously unknown corners of the galaxy. Even within the space of a few decades, technology has come far enough to provide a completely unprecedented level of detail. As the breadth of space exploration has increased, so too has the amount of data generated; and so these days, astronomers spend much time studying and interpreting information on computers.

Despite this, the romance and mystery of the skies remains intact. After centuries of watching and wondering, humankind is now finally able to venture out into the cosmos, exploring worlds that our forebears could only have imagined. Although it can't provide insights into the future or the will of the gods, studying the skies offers us a different kind of magic. Astronomy can help us to explore the outer reaches of our reality, to stretch the limits of the imagination and to boldly go out into the great mystery that surrounds us in search of answers. Nothing is more magical, nor more human, than that.

CHAPTER 9
CHEMISTRY

'Every aspect of the world today – even politics and international relations – is affected by chemistry.'

– LINUS PAULING (1901–94), SCIENTIST, EDUCATOR AND PEACE ACTIVIST

Chemistry is the science of 'stuff'. Chemists are interested in what matter is made up of, its characteristics and how it interacts with other types of matter. Often called 'the central science', chemistry intersects with many other different fields, from biology to physics. That's because chemical reactions take place all around us, every day. From your computer, to your car, to your own body, chemistry is at the heart of our lives. Learning how to master those chemical reactions has helped us to construct the world as we know it. In the present day, we use our understanding of chemistry to create things – like medicines or technology – and to understand the workings of biological life and the physical universe. But humanity has been exploiting chemistry since the earliest days of civilization – from creating pottery and tanning animal hides, to creating tools and dying clothes. Over time, we've learned to master increasingly complex chemical processes and unpick the intricate science behind the reactions. This is the story of how chemistry became fundamental to scientific enquiry, and helped to transform civilization in the process.

c. 1200 BC	*The first known chemist, Tapputi-Belatekallim, commemorated in Babylonian cuneiform tablet*
c. 5th CENTURY BC	*Empedocles proposes theory of four elements: earth, air, fire and water*

c. 9th CENTURY	*Jabir ibn Hayyan introduces early theory of materials classification*
1661	*Robert Boyle publishes the first chemistry textbook*
1702	*Georg Stahl proposes existence of phlogiston*
1783	*Antoine Lavoisier launches attack on phlogiston theory*
1803	*John Dalton demonstrates that each element is composed of specific atoms*
1869	*Dmitri Mendeleev develops modern periodic table*
1903	*Marie and Pierre Curie receive Nobel Prize for discovery of radioactivity*

THE ROOTS OF CHEMISTRY

Chemistry has always been central to the human story, right from the beginning. In the early days of humanity, heating materials over fire helped people to introduce more nutritional foods into their diet, and to create pottery. It was a growing understanding of chemical reactions that helped humans to pass from the Stone Age through to the Bronze and Iron Ages. Learning how to extract metals from their ores and to combine them to create different materials helped early civilizations to improve their tools, weaponry and infrastructure.

The first recorded chemist is Tapputi-Belatekallim, a Babylonian perfume maker whose work is described in a cuneiform tablet dating back to 1200 BC. Tapputi was an overseer at the royal palace, and she prepared mixtures for medicine and religious rituals, as well as for scent, by distilling and filtering her ingredients. Tapputi is the first known person to practise distilling ingredients in a still – a device that is still in use today. Chemistry thrived in Babylon, and the civilization's inhabitants were skilled craftspeople, adept at using chemical processes to create products such as medicines, ceramics and glass. But these inventions were generated through trial and error rather than through experimentation.

Metallurgy is the science of metals, and the study of their characteristics and behaviours.

Some of the most influential early theoretical chemistry came from Ancient Greece. From as early as 600 BC, philosophers began to form theories about the chemical makeup of the natural world. In the 5th century BC, a Pre-Socratic philosopher named Empedocles set out a theory that would go on to shape the field of chemistry for hundreds of years. Empedocles proposed that the world was made up of four fundamental elements – earth, air, fire and water – and that nothing new is ever created or destroyed, but rather transformed into a different substance by changes in the combination of these four elements.

Cuneiform is an ancient writing system developed by the Sumerian civilization of Mesopotamia in approximately the 4th millennium BC.

Empedocles' theory was adopted by Aristotle some years later, who added a fifth element – aether – thought to make up most of the universe outside the Earth. Although it was by no means the first or only theory of its kind, the Aristotelian theory would go on to shape the history of both chemistry and science as a whole.

Did You Know?

The Ancient Greeks' five-element theory of the world (earth, air, fire, water and aether) can be seen in various forms in other civilizations throughout the world. In Japan, Babylonia and India, thinkers developed similar theories, while in China a complex system called Wu Xing outlined five elements thought to govern many different areas of life.

In the early days of its history, craftspeople and philosophers helped to establish the roots of chemical thought, but it was alchemy that propelled chemistry forwards into the scientific sphere. A protoscience, alchemy

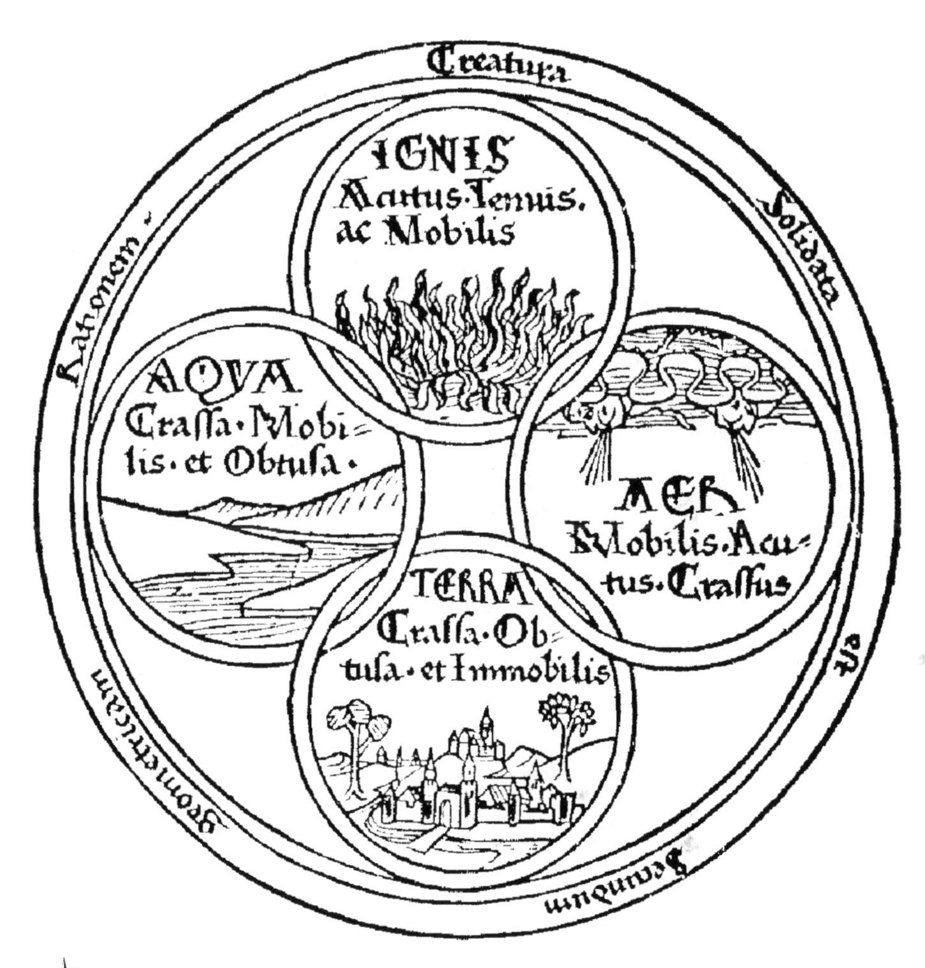

Empedocles' four fundamental elements: earth, air, fire and water.

is the quest to purify certain materials – most often by transforming metals into gold – and it combines elements of mysticism, philosophy and science. Alchemy was widely practised all over the world for many hundreds of years, and was often seen as the same discipline as chemistry, right up until the 18th century. The Aristotelian notion that one substance could be transformed into another helped to bolster alchemy in the West, and fuel its growth in the Greek and Roman periods. For centuries, scholars searched for the 'philosopher's stone': a fabled material capable of transforming base metals into gold. In the Middle Ages, alchemy and chemistry prospered in the Islamic world.

The 9th-century Persian-Arab alchemist Jabir ibn Hayyan introduced a more systematic approach to chemistry, conducting methodological laboratory experiments. Jabir was a prolific researcher – his wide-ranging studies helped to classify new chemical processes, substances and interactions, and he developed a number of experimental devices to support his research. Jabir grouped materials into three categories: metals, 'spirits', which would turn to vapour when heated, and non-malleable substances like stones, which would turn to powder when crushed. This was an early precursor to our modern system of materials classification, which includes metals, non-metals and volatile substances. Jabir's work went on to inspire medieval and Renaissance chemists in Europe and laid the foundations for the surge in research that followed. By the early modern period, the chemical sciences were in full swing throughout Europe, and over the course of the centuries that followed, they would undergo a revolution.

CHEMISTRY IN THE EARLY MODERN PERIOD

The Renaissance was a profitable, if dangerous, time for alchemists. Rich patrons funded their research, hoping to uncover the secrets of unlimited wealth and eternal life. But, of course, alchemists were not able to deliver on these investments – many earned a reputation for being fraudsters and charlatans, and some were even executed when their funders grew impatient. But while alchemy never delivered on its mystical promises, the efforts of Renaissance alchemists continued to boost the science of chemistry.

The title of first modern chemist is generally bestowed on an Anglo-Irish philosopher and intellectual named Robert Boyle. Writing in the 17th century, Boyle published *The Sceptical Chymist*, a treatise that is now generally considered to be the first chemistry textbook. In it, he criticized the Aristotelian five-element model, arguing that elements should be considered 'perfectly unmingled bodies', that is, basic substances that are the primary building blocks of matter and cannot be broken down into separate parts; a view that is much closer to how we perceive elements today. Boyle's work also presented a hypothesis that matter is made up of atoms, which he called 'corpuscles', and that all observed phenomena

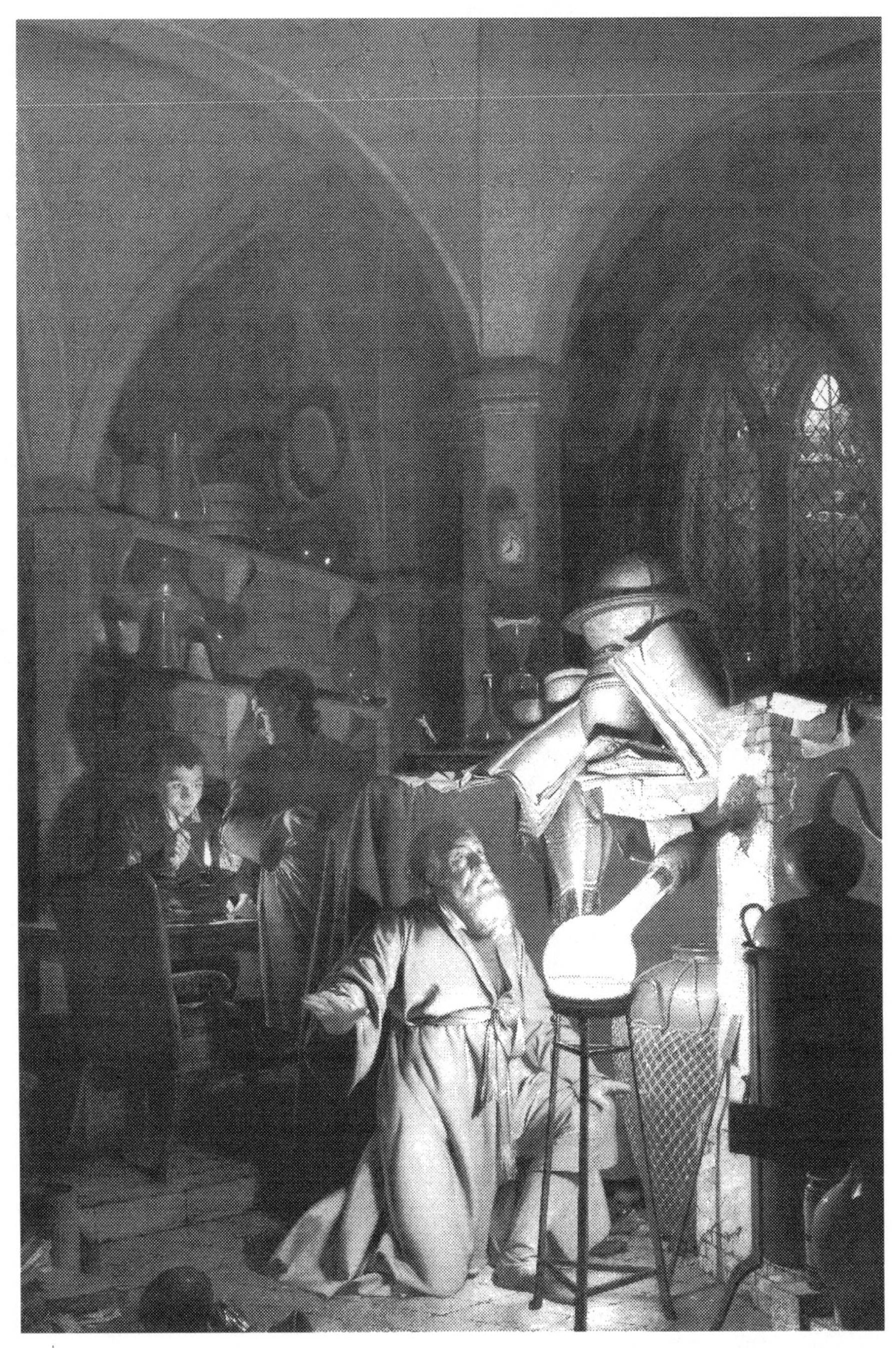

Alchemy, as depicted in this painting, The Alchymist, *by Joseph Wright of Derby, was practised all over the world for centuries, and was seen as interchangable with chemisty.*

were the result of collisions between these corpuscles. Perhaps most importantly, Boyle advocated the importance of experimentation in chemistry. Despite being interested in alchemy himself, he wanted to see chemistry recognized as an independent science, and realized that rigorous experimentation would counteract the mysticism associated with alchemy. Boyle's scientific approach was one of the first steps towards a revolution that would see chemistry enter into a new era; one marked by fact and experimentation.

By the 1700s, Boyle had been found correct in his assertion that there were many more elements than the five that Aristotle had originally proposed. In 1754, a Scottish chemist named Joseph Black isolated carbon dioxide, and two decades later, the English chemist Joseph Priestley isolated oxygen, proving that air is made up of a number of different elements and not just a single one. Priestley called his discovery 'dephlogisticated air'.

Some decades prior, in 1702, a German physician named Georg Stahl had proposed the existence of an element called 'phlogiston'. Stahl suggested that all combustible materials contained phlogiston and that it was released during the process of burning. The theory made sense to early chemists, who noted that when a candle is covered by a jar, the flame is extinguished. They surmised that this was due to an excess of phlogiston being released, causing the air to become filled with the element until the flame no longer had space to exist.

It wasn't until the end of the 18th century that phlogiston theory was finally disproven; the theory would meet its match at the hands of a French chemist named Antoine Lavoisier. Unlike his contemporaries, Lavoisier was not convinced that phlogiston existed. Through a series of experiments, he deduced that no new element was being created during the process of combustion – there was no phlogiston. What was happening, Lavoisier realized, was that an existing element was being absorbed by the combustion reaction; that element was fuelling the fire, and was eaten up by it until there was no more left, and so the flame went out. This element was Joseph Priestley's 'dephlogisticated air' or, as Lavoisier called it, 'oxygen'.

Having discredited phlogiston theory and given a new name to

 *Robert Boyle is often referred to as the first modern chemist, for his pioneering enquiry into natural phenomena.*

Priestley's element, Lavoisier went on to develop an entirely new nomenclature for chemistry, establishing many of the names and descriptions we now know today. Lavoisier's work was the culmination of the chemical revolution. By the dawn of the 19th century, chemistry had become established as its own scientific discipline, grounded in measurable fact.

Who's Who – Antoine Lavoisier

Antoine Lavoisier was born in 1743, at the dawn of the French Enlightenment – a golden age for intellectual thought in Europe. He studied chemistry at university, but found that the discipline was, at this time, ill-defined and not particularly rigorous. Over the years that followed, Lavoisier's contributions to the field would change that, leading to a revolution that ushered in a new era of chemistry.

Unlike many of his predecessors, Lavoisier was committed to taking meticulous notes and measurements in his chemical experiments. By weighing and analysing all of the substances involved in the chemical reactions he studied, Lavoisier was able to demonstrate that matter is neither created nor destroyed during these reactions. No matter how the matter is rearranged, the amount of that object remains the same – so while a gas may become a liquid, and a liquid a solid, the total mass remains constant. We now know this, the principle of the conservation of mass, as 'Lavoisier's law'.

Lavoisier's commitment to systematic chemistry was also reflected in his attempts to create a new nomenclature for the field. Up until this time, different chemists often used various terms for the same chemical compounds, making it difficult for chemists to communicate their findings to one another and build upon that knowledge. In the 1780s, Lavoisier laid out a new naming system that ensured consistency and allowed a new, more collaborative approach to chemical research.

Lavoisier made countless other contributions to the field; from dismantling phlogiston theory to creating the first modern chemistry textbook, it's hard to overstate the impact of his work. In 1794, Lavoisier became a victim of the French Revolution, when he fell foul of the revolutionaries and was executed by guillotine. Although he was only 50 when he was killed, Lavoisier had already made an indelible mark on his field. Through his commitment to consistency, measurement and reason, Lavoisier shaped chemistry into something more robust, more expansive, and altogether more modern.

EARLY MODERN CHEMISTRY

Having established the rules and principles that governed the science of chemistry, scholars in the 19th century began building on those foundations. In the early 1800s, an English teacher and scholar named John Dalton determined the existence of atoms, reviving the assertion that Ancient Greek philosopher Democritus had made more than 2,000 years before. But Dalton went one step further, arguing that each element is composed solely of its own unique type of atoms, and that these atoms vary in size, weight and properties. Of course, Dalton couldn't just weigh elements' atoms to prove this, so what he did instead was to assess the weight of each element in relation to the others. In this way, Dalton was able to determine that hydrogen atoms are lighter than carbon, and so on. Dalton also worked out how atoms from different elements can combine in set ways to form chemical compounds, such as carbon dioxide. He noted that chemical reactions can lead to atoms becoming rearranged into compounds, but the atoms themselves are never created or destroyed.

Did You Know?

Before Jöns Jacob Berzelius established the practice of referring to all chemical elements by letters, some were represented with symbols, such as a circle with a line through it or a plus sign inside.

John Dalton published his atomic theory in 1808.

Shortly after this, a Swedish chemist named Jöns Jacob Berzelius used Dalton's atomic theory to create a table of elements, with the order based on their atomic weights relative to one another. Building on this system, Berzelius invented a system of naming the various elements; each was marked by one or two letters – for instance, H for hydrogen, Mg for magnesium. In 1869, Dmitri Mendeleev, a Russian chemist, took Berzelius' table and completely flipped it, so that the elements were arranged horizontally in order of their atomic weight and placed into vertical columns containing elements with similar properties. Mendeleev's table demonstrated a clear order to the elements, and the arrangement could even be used to identify gaps in the table, where presently undiscovered elements should sit. After these missing elements – gallium, scandium and germanium – were discovered in the 1870s and 1880s, it became clear than Mendeleev's table had cracked the code and captured the inherent order within the chemical elements.

The Voltaic Pile

The Italian physicist Alessandro Volta was the first to invent a functional electrical battery – a discovery that helped to usher in a new age of research into electrical phenomena. Working at the turn of the 19th century, Volta was intrigued by the concept of 'animal electricity'. In the 1790s, another Italian scientist, Luigi Galvani, had observed that movement could be triggered in the legs of a recently dead frog if the nerves were connected to two different metals.

Galvani concluded that the movement resulted from 'animal electricity' – an inherent form of electricity inside living tissue. Volta, however, was not convinced. He believed that the frog's legs were just acting as the conductor of an electric current flowing between the two metals. To demonstrate this, Volta created a device to experiment with the mysterious force. In 1794, he made a circuit of two metals separated by a piece of brine-soaked cloth. With this circuit, Volta was able to show that the animal tissue Galvani had claimed was

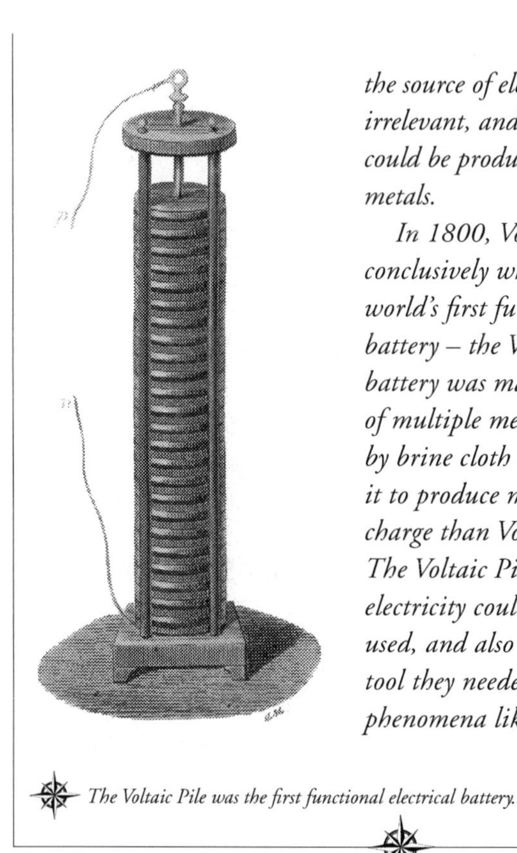

the source of electrical charge was irrelevant, and that electrical current could be produced with only the metals.

In 1800, Volta established this conclusively when he produced the world's first functional electrical battery – the Voltaic Pile. The battery was made up of stacked pairs of multiple metallic discs, separated by brine cloth – a set-up that allowed it to produce much more electrical charge than Volta's previous circuit. The Voltaic Pile demonstrated that electricity could be harnessed and used, and also gave scientists the tool they needed to study electrical phenomena like never before.

The Voltaic Pile was the first functional electrical battery.

Dalton's atomic theory had become a cornerstone for chemistry, and it was clear that chemistry was beginning to intersect more closely with the science of physics. By the end of the 19th century, the new subfield of physical chemistry had begun to grow, with experiments into electrochemistry and thermodynamics helping to establish the importance of the field. One of physical chemistry's most accomplished practitioners was Marie Sklodowska-Curie, a Polish physicist and chemist. Experimenting initially on uranium, Curie and her husband Pierre discovered that the element was responsible for emitting strange rays – they called the phenomenon 'radioactivity'. In the years that followed, the Curies identified two further radioactive elements: polonium and radium. From nuclear energy to radiotherapy, the vast significance of

Marie and Pierre Curie at work in their laboratory in Paris, where they discovered radioactive elements.

these new elements in energy, medicine and technology would become apparent over the course of the 20th century.

In modern times, chemistry has become even more closely connected to both physics and biology. An understanding of the basic tenets of chemistry is fundamental to virtually any aspiring scientist, because of its centrality to all areas of research. But chemistry is not only crucial to scientific research, it's also completely embedded in our daily lives. Our understanding of chemical reactions has shaped the world in which we live today. Chemistry is all around us, all the time, providing the very foundations for life as we know it.

PART FOUR

THE 19TH CENTURY

THE 19TH CENTURY was an exciting time to be a scientist. Over the course of 100 years, science became established as a profession, distinct disciplines emerged and research grew increasingly specialized. A series of major discoveries – from evolution to electromagnetism – added to the palpable sense of scientific momentum.

Meanwhile, the wider world was also in flux. The new age of European imperialism saw a resurgence of colonization, particularly in Africa, as well as Asia and the Middle East. The exploitation of colonies brought political and economic benefits to colonizing nations, usually at the expense of the colonized lands. Amidst all of this, science was seen as a source of national pride and an important tool for social progress.

At the same time, the Industrial Revolution transformed the landscapes of countries like Britain, France, Germany and the US, leading to increased urbanisation and population growth. The drive towards industrialization created new opportunities for scientists and drove a spate of new inventions and ideas.

The importance of research to society and national prestige supported the professionalisation of science, and working as a scientist became a well-respected and recognized career choice. What's more, increased specialization in the sciences and the branching off of science into separate disciplines supported the emergence of professional bodies, such as the Royal Institution in Britain, and the American Association for the Advancement of Science in the US.

The 19th century was a time of optimism and achievement in the sciences, and the push to answer increasingly complex questions led to unprecedented advances in both theory and practice across a range of different fields. By the close of the century, scientists had helped to usher in the industrialized world that we know today. In doing so, the sciences had become established as fundamental to national progress and a cornerstone of modern society.

CHAPTER 10
ENGINEERING

'The engineer has been, and is, a maker of history.'
– JAMES KIP FINCH (1883–1967), ENGINEER AND EDUCATOR

Engineering can be thought of as science in action. Engineers use scientific knowledge to figure out how things work, solve practical problems, design things and generate innovations. When we think of engineers, we might think of people who design bridges or cars, but the field goes far beyond that, spanning areas of biology, chemistry and physics. There are four main categories of engineering: civil, mechanical, electrical and chemical. So, while some engineers might work on creating artificial limbs, others might design robots; some could work on improving manufacturing processes, while others will work to combat the impact of air pollution on the environment. All of these types of engineers form part of a long and fascinating history; human beings have been innovating for thousands of years. From the invention of the wheel to the creation of smartphones, the spirit of human ingenuity is the thread that weaves the history of engineering together.

c. 2600 BC	*Imhotep designs Pyramid of Djoser*
c. 3rd CENTURY BC	*Archimedes designs Archimedes' Screw*
1687	*Isaac Newton establishes classical mechanics*
1799	*Alessandro Volta invents the electrical battery*
1816	*Francis Ronalds designs first working electrical telegraph*
1885	*Karl Benz invents the 'Motorwagen'*

1895 *Wilhelm Röntgen discovers X-rays*

1903 *Wright Brothers complete their first successful flights*

1928 *Rolf Wideröe builds first linear accelerator*

1958 *Tom Brown develops prototype for a medical ultrasound machine*

THE ROOTS OF ENGINEERING

One of the earliest outlets for engineering ingenuity was through what we would now recognize as civil engineering. From ancient Mesopotamian aqueducts to the Colosseum in Rome, the knowledge and skill of ancient engineers is still evident today. The first official civil engineer was an Ancient Egyptian government official named Imhotep. Around 2600 BC, Imhotep designed and oversaw the construction of the Pyramid of Djoser, a limestone structure around 200 feet high – that's roughly the equivalent of a modern 18-storey building. Around a century after this, Egyptian engineers built the Pyramid of Giza. Despite working without wheels, pulleys or iron tools, the engineers built this pyramid 481 feet high – and it remained the tallest manmade structure in the world for nearly 4,000 years.

The Romans were also famed for their engineering. As part of their Empire-building, they constructed many miles of roads, spanning Asia, Africa and Europe. Roads like the Appian Way – a 350-mile trail running south through Italy – were constructed using an early form of concrete, and designed with drainage ditches and a slight camber to prevent flooding. Other examples of Roman ingenuity can be seen in their sanitation systems – impressive aqueducts helped to carry clean water to the cities, while a network of sewers ensured hygienic waste disposal. From transportation to sanitation, the evidence of Roman architecture still stands today, as testament to the Romans' immense engineering skills.

An Archimedes' Screw is a device that draws water up through a cylinder using a rotating blade.

While demand for a high-functioning infrastructure spurred advances in engineering, another driver of progress was conflict and war. For instance, mechanical engineering – the design of machines – progressed in part because of the need to build new and more deadly weapons. Prolific in the 3rd century BC, the Greek mathematician and engineer Archimedes designed catapults and ship-sinking devices, along with a pulley system that allowed for the movement of heavy loads. Away from weaponry, Archimedes also used his engineering knowledge to invent remarkable machines such as the Archimedes' Screw, which is still used in industry today.

Did You Know?

Mechanical engineering also prospered in China, where thinkers invented the seismometer, geared chariots and hydraulic-powered bellows.

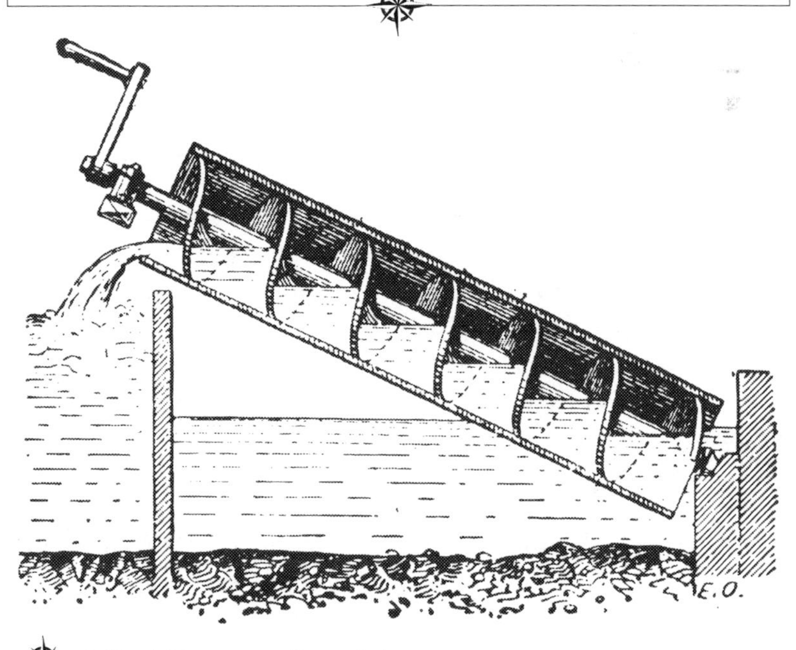

Archimedes' Screw: a pump for transferring water.

While engineering innovations did occur throughout the Middle Ages, particularly in the Islamic world, it was in the last years of the 17th century that the theoretical foundations of modern engineering began to emerge. In 1687, English polymath Isaac Newton published *Principia*, his scientific masterpiece, which included the revolutionary laws of motion. These laws described the physical principles behind why and how objects move, and helped to lay the foundation for classical mechanics: the physics of motion. With its grounding in mathematics, classical mechanics helped to provide a guiding theory for engineers of all backgrounds.

With the theoretical foundations in place, the 18th century saw the rise of engineering as an independent profession. At the same time, both civil and mechanical engineering surged forwards when the Industrial Revolution began to get underway in Britain. By the end of the century, life was beginning to change drastically, as industrial innovation started to reshape the landscape and transform wider society. Over the course of the next century, the effects of industrialization would spread to other countries, ushering in a new era of engineering and transforming nations around the world.

ENGINEERING IN THE 19TH CENTURY

The 19th century saw huge changes in everyday lives and landscapes. In many countries – including the UK and US – vast numbers of people moved from the countryside to the cities, and as the urban environment grew and industry expanded, so too did opportunities for engineers. Civil engineers, such as Isambard Kingdom Brunel in Britain, Mokshagundam Visvesvaraya in India and Benjamin Wright in the US, continued to develop ever grander and more innovative infrastructure projects. Meanwhile, mechanical engineers set to work on designing increasingly complex machines. In the 1830s, the English mathematician Charles Babbage designed his analytical engine: a mechanical precursor to the computer, capable of performing complex calculations. The century also saw the invention of the internal combustion engine, the steam train and the motor car. But in addition to advances in mechanical and civil engineering, new disciplines also began to emerge.

Over the centuries, engineers developed ever grander and more innovative infrastructure projects. The Krishna Raja Sagara (KRS) Dam was built in 1911 to combat regular droughts in the Mysore region of India.

Who's Who – Isambard Kingdom Brunel

One of the most widely renowned figures in engineering history, Isambard Kingdom Brunel is also remembered as one of the most innovative and visionary thinkers of the 19th century.

Brunel was born into engineering – his father, Sir Marc Isambard Brunel, was also highly regarded in the field, and it was through this family connection that Brunel began work on the Thames Tunnel in England in 1825. The tunnel was the first of its kind and, running at 23 metres deep and 300 metres long, it was an astonishing feat of engineering for its time.

After the Thames Tunnel, Brunel began working on a series of ever more ambitious bridges. Part of what made Brunel such a successful engineer was the scale of his ambition. In 1833, when Brunel was appointed chief engineer of the Great Western Railway, he began drawing up plans to extend the railway all the way to New

York. Brunel envisioned passengers changing from their steam train to a steamship in Wales, and sailing the rest of the way across the Atlantic.

Brunel went on to design a series of steamships, the first of which proved significantly more efficient than any of its predecessors. Named the Great Western after the railway Brunel had built, the ship became the first steamship to provide a regular service across the Atlantic.

Brunel's first project, the Thames Tunnel, is now part of the London Overground, while many bridges – including the iconic Clifton Suspension Bridge in Bristol – remain in use today. From ships to dockyards, bridges to railways, Brunel's designs have stood the test of time and left a lasting mark on the British landscape.

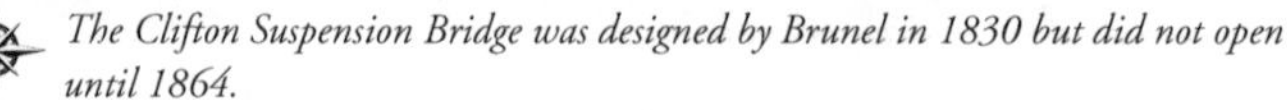

The Clifton Suspension Bridge was designed by Brunel in 1830 but did not open until 1864.

Did You Know?

London's existing sewerage network was completed back in the 1870s. Prior to this, human waste had been disposed of in the River Thames, leading to terrible pollution. Designed by Sir Joseph William Bazalgette, the introduction of an enclosed system of waste disposal improved sanitation and dramatically reduced the incidence of diseases such as cholera.

The Industrial Revolution created a surge in the demand for chemical engineers: experts who specialize in the creation of products using chemical processes. In previous years, products were mainly produced through batch processing – in which mixtures are prepared in discrete batches – or through cottage industries: businesses run out of people's homes. But during the 19th century, manufacturers strove to improve

Karl Benz at the tiller of the improved model of his 'Motorwagen', offered for sale in 1887.

efficiency and increase the amount of goods they could produce. Chemical engineers were at the forefront of this shift towards a new era of mass production, designing chemical plants and processes that allowed for the continuous production of goods. Meanwhile, in 1885, a German engineer named Karl Benz invented the first automobile – the 'Motorwagen'. But these new machines needed a key chemical in order to run: petroleum. The process of extracting useful fuels from crude oil soon grew into a major industry, and large oil refineries were built all around the world. Managing complex chemical processes on huge scales required specialized engineering knowledge, and by the end of the century, chemical engineering had become fully embedded in the industrial economy.

The Industrial Revolution

The advent of modern engineering was closely tied to the Industrial Revolution which took place between the mid-18th and mid-19th centuries, first in the UK and then in other countries around the world. This was a period of change, marked by significant progress in manufacturing methods, an abundance of new products and stark changes to the industrial landscape.

But this period in history consisted of much more than advances in engineering and technology; it was also a time of rapid socio-economic and cultural change. By the end of the 19th century, everyday life had been transformed. Increased urbanisation, mass production, changes to labour and advances in transport and communication systems all impacted on both the individual and society at large.

A key facet of the Industrial Revolution was the drive for greater efficiency, and advanced technology and engineering proved key to modernising industry. In 1712, an English inventor named Thomas Newcomen invented the first commercially viable steam engine. He created his engine – the Miner's Friend – with a view to pumping water out of Britain's coal mines. Some years later, in

1781, the Scottish engineer James Watt improved on Newcomen's model, doubling its efficiency. Alongside advances in iron production and textile manufacturing, the invention of the steam engine was one of the major developments in the history of industrialization. The improved engine provided industry with a more efficient and reliable source of energy, and it could be used to power a huge range of industrial processes.

From Britain, the Industrial Revolution spread to Belgium, France, Germany, the US and Japan, transforming economies and communities as it went. There is debate about the extent to which the living standards of working people improved as a result of increased industrialization. There were marked downsides to the revolution, including poor labour conditions and cramped, unhygienic living quarters. But what is clear is that the Industrial Revolution was central to shaping the world as we know it today. For both science and society, nothing would ever be the same again.

But chemical engineers weren't the only faction to prosper during the 19th century. Electrical engineering is concerned with the design and function of electrical systems, such as power lines or telecommunications. Prior to the 19th century, scholars were aware of electrical phenomena, but the transformative potential of electricity had yet to be realized. At the turn of the century, in 1799, the Italian physicist Alessandro Volta invented the world's first electrical battery (see page 135). Just 17 years later, a young English inventor named Francis Ronalds designed the first working electrical telegraph. At first, his invention was ridiculed by the governing elite, who deemed it unnecessary. But by the end of the century, telegraph systems spanned countries and continents, connecting previously disparate areas of the world for the first time. From the incandescent lamp to the telephone, by the close of the century, electricity had gone from an obscure phenomenon to a household commodity. And as the demand for electrical appliances and innovations grew, so too did the work of electrical engineers.

THE ROAD TO MODERN ENGINEERING

The 19th century was characterized by the growth and diversification of engineering, and that trend continued right on through to the 20th century. In modern times, the remit of engineers has grown even larger, and more specialist disciplines have continued to emerge.

Did You Know?

When Wilhelm Röntgen discovered his mysterious X-rays, he used them to take a radiograph image of his wife Bertha's hand. Poor Bertha was so unnerved by the experience that she reportedly exclaimed: 'I have seen my death!'

The discovery of X-rays in 1895 was one of the early steps towards a new era of biomedical engineering. First described by a German mechanical engineer and physicist named Wilhelm Röntgen, X-rays are high-energy beams capable of passing through matter that normal light cannot, such as skin, fat and muscle. Röntgen's discovery triggered a spate of new medical inventions designed to exploit the diagnostic and curative powers of X-rays and other forms of high-energy radiation. In 1928, a Norwegian physicist named Rolf Widerøe built the world's first linear accelerator: a type of particle accelerator that could create a beam of high-speed subatomic particles. Biomedical engineers soon used Widerøe's invention to develop a machine for use in cancer treatment, and in 1953, the device was used to deliver the first radiotherapy treatment.

During the latter half of the 20th century, advances in computing greatly expanded the possibilities for biomedical engineers. The invention of transistors by Bell Laboratories in New York in 1948 allowed engineers to create a new generation of clinical implants and equipment. From hearing aids to pacemakers, medical devices that relied on electricity became much smaller and more efficient. Computer techniques also aided the development of new devices to support medical imaging. In the 1950s, a young Scottish engineer named Tom Brown developed the first prototype for a medical ultrasound machine. The device paired a probe, which sends and receives sound waves, with a computer processing unit to interpret the waves and create an image of the internal human body. Along with implants and devices, ultrasound machines were just one of the many biomedical innovations spurred on by advances in electronics and computing.

Did You Know?
Transistors are tiny electronic components that control the flow of electricity in electronic devices such as computers.

The advance of engineering has always been closely tied to historical events. In the 20th century, a series of major wars spurred swift progress in the field of aerospace engineering: a field that encompasses both aircraft and spacecraft. The successful flights of the Wright Brothers occurred in 1903; and by the 1910s, engineers had begun working on the

Bell Lab scientists who invented transistors (left to right): John Bardeen, William Shockley and Walter Brattain.

design of military aircraft for use in the First World War. Innovations in aerospace technology demanded a fleet of specialist engineers capable of fine-tuning every element of these complex machines. But before long, aerospace engineers would find themselves part of something even more monumental. In 1926, an American aerospace engineer named Robert Goddard completed the first successful flight of a liquid-propelled rocket. Goddard's work demonstrated that flight was possible at speeds greater than the speed of sound, opening up a wealth of new possibilities for aerospace engineers. In 1957, Sputnik I – the world's first satellite – was launched from Kazakhstan, marking a new era in space exploration, and by the end of the 1960s, human beings had set foot on the Moon for the first time. In the decades that followed, spacecraft travelled throughout the solar system and beyond; and at the centre of all these achievements were aerospace engineers. Back on Earth, from the telecommunications industry to the development of more powerful and efficient computers, aerospace engineering has helped transform the world as we know it.

The 20th century was a time of profound diversification in engineering. In addition to biomedical and aerospace applications, engineers also began to specialize in environmental science, industry, robotics and many more fields. Today, we see engineers of all backgrounds working on increasingly technical and specialized projects, but the spirit of innovation and creativity remains. The world as we know it now was built by engineers, and the future ahead of us will be too.

CHAPTER 11
GEOLOGY

'The stony science, with buried creations for its domains, and half an eternity charged with its annals....'

– HUGH MILLER (1802–56), GEOLOGIST, WRITER AND FOLKLORIST

Forget what you know; geology is so much more than just rocks. Geology encompasses an enormous range of different types of science, and its history spans the fields of physics, biology and chemistry. There are two main branches of geology: the study of the planet's physical processes and features, and the study of the planet's history. Those two things can tell us whole lot; geology has shaped the way in which humans live, and remains fundamental to our lives today. Understanding geology has many practical uses: it helps us to exploit resources and generate power, to recognize and mitigate environmental hazards, and to source materials for building infrastructure and products. But beyond its immediate practical applications, geology can also help us to understand the fundamental processes that have shaped the Earth as we know it. By studying the physical landscape of our planet, geologists can help us look backwards in time and chart the history of the Earth and its inhabitants.

c. 4th CENTURY BC	*Aristotle proposes idea of slow geological change*
1074	*Shen Kuo proposes early theory of geomorphology*
1666	*Nicolas Steno founds stratigraphy and theorizes existence of fossils*
1741	*Georges-Louis Leclerc suggests the Earth is much older than widely thought*

1815	*William Smith publishes first geological map of England and Wales*
1830	*Charles Lyell popularizes theory of uniformitarianism*
1907	*Bertram Boltwood uses radiometric dating to estimate the age of the Earth*
1915	*Alfred Wegener proposes theory of continental drift*
1953	*Marie Tharp discovers the Great Global Rift*
1960	*Harry Hess describes the mechanism behind continental drift*

THE ROOTS OF GEOLOGY

Long before there was any way of knowing how old the Earth was, thinkers around the world were asking that very question. In the 4th century BC, Aristotle proposed that the Earth's structure and landscapes had changed very gradually over an extremely long time. Some centuries later, in Song dynasty China (960–1279), the polymath Shen Kuo observed the presence of marine fossils hundreds of miles inland, and proposed that the seashore had shifted and the land had become reshaped over many years. This is one of the earliest known theories of geomorphology, and a strikingly accurate account of land formation.

During the 15th and 16th centuries, people thought that fossils resulted from the Earth trying to grow new life inside it. But the Renaissance polymath Leonardo da Vinci, who came from Tuscany, a fossil-rich region of northern Italy, believed that fossils could be the remains of life that had previously existed. Like Shen Kuo before him, Leonardo suggested that regions containing fossils resembling sea creatures were probably once covered by the sea, and that the movement of water had shaped the landscape over huge periods of time.

In geology, sediment is solid matter that consists of soil, rock or organic remains that have been moved and deposited elsewhere by erosion.

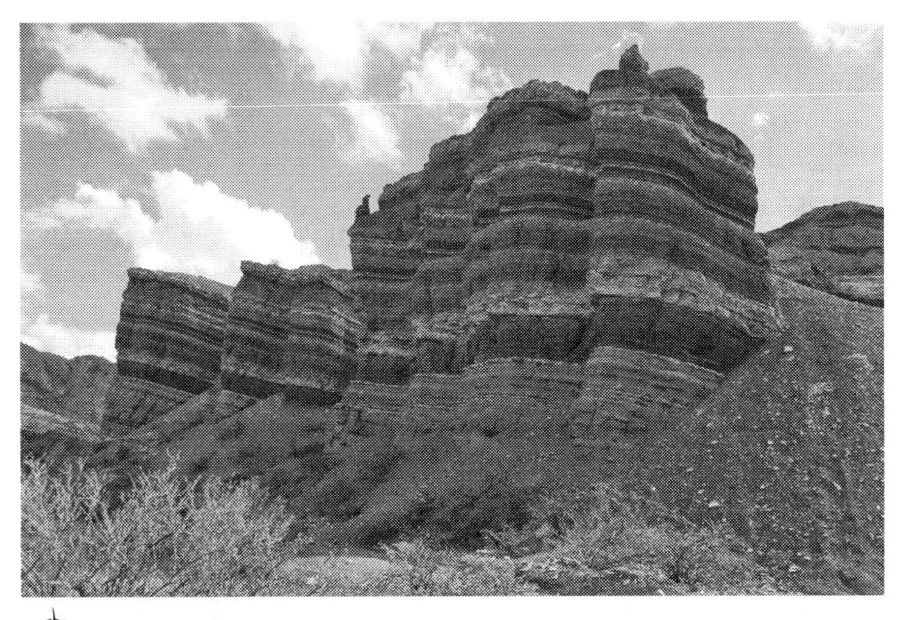

Sedimentary rocks, such as these in Cafayate in Argentina, display the Earth's long geological history.

The 17th century saw debate grow over the age of the Earth, spurring an increase in geological research that, in turn, prompted a number of new discoveries. While theologians often argued that the Earth was only as old as suggested in the Bible, evidence began to mount that the planet was, in fact, much older. In 1666, a Danish scientist named Nicolas Steno was studying the head of a great white shark when he noticed that the animal's teeth looked remarkably like 'tongue stones': objects that were often found embedded in rocks. Steno theorized that the tongue stones were, in fact, ancient shark's teeth that had become mineralised and turned to stone over many years. He also suggested that they had ended up in rocks after being gradually buried in sediment over time. This sediment, he reasoned, settled in horizontal layers, with the oldest layers at the base and the newer layers near the top. This discovery was one of the earliest steps in the evolution of modern geology and, although Steno never publicly acknowledged it, implicitly challenged the idea that the Earth was the age that many theologians claimed. After all, processes like the one Steno was describing could only occur over lengthy periods of time.

The idea that the Earth was only a few thousand years old was directly challenged in the mid-18th century, when a French naturalist named Georges-Louis Leclerc, Comte de Buffon, demonstrated through experiments with heated iron spheres that the Earth was probably much older. By observing the time it took for the heated spheres to cool down, Buffon estimated that the planet was around 75,000 years old.

Although it was becoming clear that the Earth was much older than scholars had previously believed, the nature of geological cycles was still very much up for debate. In the 18th century, many geologists believed in the theory of 'catastrophism', which held that the planet's landscapes had been formed by a series of cataclysmic events – like the great flood mentioned in the Bible. This idea was challenged in the 1780s by Scottish geologist James Hutton, who argued that geology was characterized by a cycle of routine activity taking place over vast timescales. Rather than being shaped by devastating floods or other disasters, Hutton posited that the Earth's geology had been formed by regular and predictable processes. This theory came to be known as 'uniformitarianism', and it would be hotly debated in the years to come.

Diluvialism

In the Bible, the Book of Genesis describes a great flood sent by God to cleanse the world of its sins. According to this account, the flood covered every part of the Earth, and the only survivors were God's faithful servant Noah and his family, who boarded an ark, along with two of every creature.

During the 17th century, geology grew rapidly, but the science underpinning the field also faced challenges from religious thought. In Europe, many people believed that the global flood described in Genesis had actually taken place and was the basis for real geological phenomena. This theory became known as diluvialism.

When the fossilized remains of sea creatures were found inland, some thinkers took this as evidence that a great global flood had taken place. We know today, however, that this phenomenon results from the changing geomorphology of the landscape over many millions of years. As geological research progressed and it became clear that the Earth was much older than Biblical chronology suggested, the notion that the flood was a real event began to fall out of favour. While some geologists, known as Neptunists, continued to argue that geological features were caused by a series of floods, the theological argument – with its emphasis on an improbably young Earth – began to wane in the 18th century.

By the mid-19th century, diluvialism had mostly fallen out of favour. Charles Darwin's On the Origin of Species, *published in 1859, challenged the account of Earth's creation as laid out in Genesis, and further undermined the credibility of the Bible as an accurate source of geological history.*

We now know that, while the Earth may have experienced massive floods in its history, the events described in Genesis are by no means a reflection of historical fact. In fact, the forces of heat and pressure have been the most significant in determining the geological formations we see around us. These processes have occurred over billions of years – indeed, our planet's history is far longer and far more complex than our ancestors could ever have imagined.

19TH-CENTURY GEOLOGY

The 19th century was a time of rapid growth for geology. In Britain, the Industrial Revolution increased the demand for coal, which in turn spurred the desire for knowledge about the subsurface and composition of the Earth. In 1815, a British engineer named William Smith published the first geological map of England and Wales, which depicted 23 different kinds of strata represented in different colours. Having explored Britain's strata in such meticulous detail, Smith realized that each layer of rock contained fossils that could not be found in any other layer.

On the other side of the Channel, French zoologist Georges Cuvier made a similar observation. Cuvier was fascinated by fossils, and he recognized that they could be used to work out the relative age of different strata. Cuvier found that the most exotic and unrecognizable fossils were buried within the deepest, oldest layers of strata; he concluded that these fossils did not belong to any modern species of creature, but were from species that had since gone extinct.

Strata are layers of soil laid down by sediment, often in the sea. They are slowly changed into rock by pressure, heat and chemical reaction.

The idea that entire species of animals had gone extinct gave weight to the theory of catastrophism, which remained popular until the 1830s, when Charles Lyell, a Scottish geologist, published his *Principles of Geology*. In this seminal text, Lyell argued against catastrophism, suggesting that geologists could understand the history of the planet's geology simply by looking at processes currently taking place around them. Changes occur very slowly, Lyell argued, but they occur in standard, predictable ways, not through natural disasters. Lyell's work was highly successful, and triggered the downfall of catastrophism. Even today, while scientists note the geological significance of one-off cataclysms, uniformitarianism is still recognized as the presiding geological theory.

Who's Who – Mary Anning

On the shore of south-west England, there is a stretch of coastline almost 100 miles long, known as the Jurassic Coast. A recognized World Heritage Site, this coastline contains rocks spanning 185 million years of geological history, from as early at the Triassic Period. As such, the Jurassic Coast is rich with fossils. It was in this geologically significant region that palaeontologist Mary Anning was born in 1799. Anning grew up without the means that many of her scientific contemporaries enjoyed. Her family was poor, and like many girls at the time, Anning did not receive much in the way of formal education.

William Smith's pioneering geological map of England and Wales was first published in 1815.

When she was just 11, Anning's father died, leaving the family heavily in debt. So Anning set to work in order to support her mother and brother. Following in her father's footsteps, Anning became a fossil collector, selling the specimens she found on the

Mary Anning is now regarded as a hero of paleontology.

coastline to scientists and other enthusiasts. At the age of 12, Anning and her younger brother discovered an ichthyosaur skeleton: a large marine reptile that first emerged up to 250 million years ago. Some years later, Anning uncovered the first complete plesiosaurus skeleton, belonging to another large marine reptile.

As time went on, Anning began to teach herself geology and anatomy, so that she could better understand the specimens she was working with. By the time she reached adulthood, Anning had become an expert in the field of palaeontology. She opened a store and supplied specimens to geologists and fossil collectors from all over Europe.

Despite her knowledge and prowess, however, Anning's achievements were overlooked and frequently written out of geological reports. The Geological Society of London refused to admit her on the grounds of her gender. But word of Anning's abilities began to spread, and she became well known in scientific circles, including to Charles Darwin, whose theory of evolution was probably influenced by Anning's work.

Although recognition did not come in her lifetime, Anning is now regarded as one of history's foremost fossil-hunters, and a giant in the field of palaeontology.

As time went on, geology began to intersect with other fields of science. In 1862, William Thomson, a Scottish engineer and physicist, applied thermodynamic principles to determine the age of the Earth, and calculated that the planet was between 20 million and 400 million years old. Thomson argued that Earth would have once been entirely made up of molten mass, and that the amount of time taken for the planet to cool to its existing temperature would have been many millions of years; this work was revolutionary in demonstrating the ways in which the principles of physics could help geologists dig deeper. Chemistry, too, began to converge with geology, as new techniques emerged to determine the elements within minerals and rocks. Attempts were made, by the Irish physicist John Joly and many others, to determine the age of the Earth based on the sodium content of the oceans. Although these estimates, too, were ultimately flawed, they demonstrated the potential applications for chemistry in the field. At the close of the 19th century, many mysteries remained, but the foundations for modern geology had been firmly established.

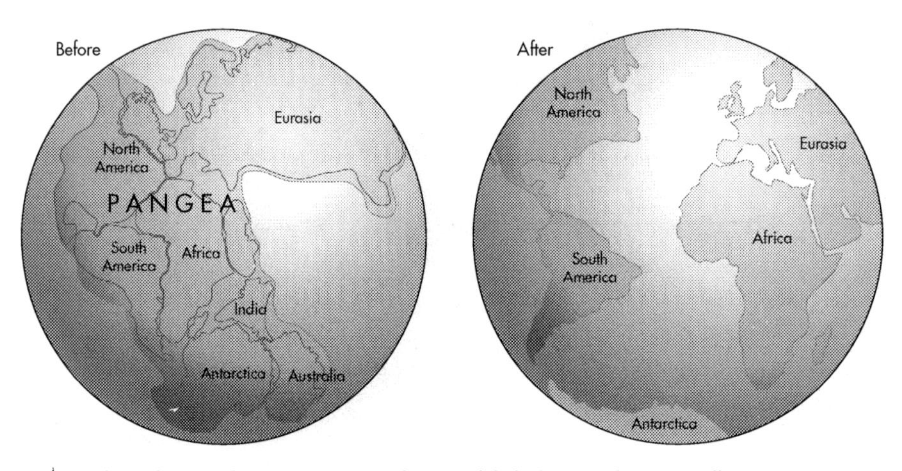

The single, original supercontinent named Pangea (left) broke apart about 175 million years ago, because of continental drift.

Did You Know?

William Thomson's 1862 estimate for the age of the Earth was more accurate than any before, thanks to his use of thermodynamic techniques. But the real figure was still way off, because Thompson failed to account for the extra heat produced by radioactive decay and convection.

THE ROAD TO MODERN GEOLOGY

In the early years of the 20th century, a game-changing technology emerged: radiometric dating. The American chemist Bertram Boltwood was a pioneer of this method, and used it to place the age of the Earth at 2.2 billion years – a huge jump from previous estimates. Within a few short years, geologists had come to accept that the Earth was, in fact, billions of years old.

Radiometric dating works by determining the levels of radioactive decay inside a material, and using known decay rates to estimate the material's age.

However, it wasn't long before a new debate began to emerge around the concept of continental drift. In 1915, German meteorologist and geophysicist Alfred Wegener published his highly controversial text, *The Origin of Continents and Oceans.* In the book, Wegener argued that all existing continents had once been joined together in a single large land mass. He called this single continent Pangea, and suggested that it had gradually broken apart over a vast period of time. This theory, named 'continental drift', was not widely accepted to begin with, and for many years it remained overlooked entirely.

Did You Know?
The name 'Pangea' to describe Alfred Wegener's proposed supercontinent comes from the Ancient Greek word pan, *meaning 'entire' or 'whole', and* Gaia, *meaning 'Mother Earth'.*

But Wegener would be vindicated some decades later, when geologists uncovered what has been hailed as the unifying theory of geology: plate tectonics. Plate tectonics is the theory that the Earth's outer shell, called the lithosphere, is made up of moving plates sitting on top of a molten rock layer, called the asthenosphere. Geological features and activity – such as mountain ranges, earthquakes and volcanic processes – occur as a result of the movement of these plates.

In 1953, a team of American physicists – Marie Tharp, Bruce Heezen and Maurice Ewing – were mapping a vast underwater mountain system

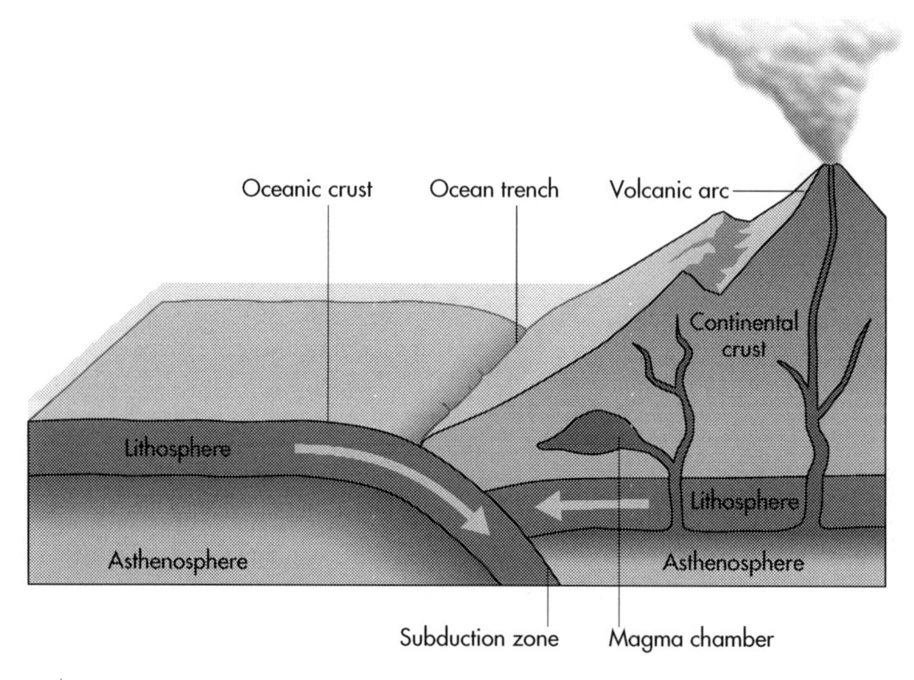

Plate tectonics, in which the Earth's outer shell is divided into several plates that glide over its mantle, is a unifying theory in geology.

called the Mid-Atlantic Ridge, when Tharp discovered a deep rift running through the ridge. Now known as the Great Global Rift, this geological feature seemed to suggest that the Earth's crust was splitting and being pushed apart in this particular spot.

In 1960, another American geologist, Harry Hess, worked out exactly what was going on. Hess proposed that the rift was caused by molten rock from the planet's mantle being pushed upwards along mid-ocean ridges. This new material was then forcing the existing crust outwards, causing the seafloor to be pushed apart, creating a rift. Hess's findings suggested that it was entirely possible for the continents to have split apart over time, and provided a mechanism for how this could take place.

Over the course of the 1960s and 1970s, it became clear that the geological activity taking place at the Mid-Atlantic Ridge was the product of plate tectonics. Research into the age of the ocean floor, patterns in rocks from the seafloor crust, and the clear correlation between volcanic activity and the proposed plate boundaries all served to create consensus in the geology community that plate tectonics theory was accurate. We now know that the Mid-Atlantic Ridge marks the boundary of one of the seven major plates that make up Earth's crust, and that the movement of these plates creates geological features such as mountain ranges and canyons, as well as natural phenomena like volcanic eruptions and earthquakes. That's why plate tectonics is considered a unifying theory for geology, because it provides an underlying explanation for so much of the planet's geological features and phenomena.

Although many great mysteries have been solved and many raging debates settled, geology continues to remain more relevant now than ever before. In the 21st century, geologists exploit a much greater range of technology – from satellite imagery to computer modelling – to provide even more powerful insights into the science of the Earth itself. In the present day, geologists are helping us to track climate change, to respond to geohazards like earthquakes, and to seek out new sources of energy and other resources. The Earth is made up of many complex systems, but thanks to geologists, we continue to unpick the secrets of the planet we call home.

CHAPTER 12
MICROBIOLOGY

'How can things so small be so important?'
– ISAAC ASIMOV (1920–92), AUTHOR AND BIOCHEMIST

Microbiology is the study of microorganisms, such as bacteria, viruses and other microscopic forms of life. Microorganisms are the most abundant type of life on the planet; it's thought that there are more microbial species on Earth than there are stars in the galaxy. Even within our own bodies, there are trillions of microbial cells, probably outnumbering our own human cells. Microbes occupy virtually every corner of the planet, and that's just as well because, from producing oxygen to disposing of environmental waste and pollution, from supporting human immune systems to providing food for animals, microbes are intimately tied to life. But microbes aren't all universally good for us; some forms of viruses, bacteria, parasites and fungi can be deeply harmful to other forms of life, including humans. From vaccines to antibiotics, learning how to manage these organisms has been a vital part of medical history. The more that we learn about microbes, the more there is to admire about them – their resilience, adaptability and abundance have allowed them to prosper like no other group of organisms. Humankind is generally considered the dominant species, but taken together, surely microbes actually dominate the Earth.

1546 *Girolamo Fracastoro suggests epidemics are caused by seed-like spores*

1676 *Antonie van Leeuwenhoek discovers bacteria*

1837 *Theodor Schwann and Charles Cagniard de La Tour demonstrate that microorganisms cause fermentation*

1865 *Louis Pasteur establishes germ theory of disease*

1879 *Pasteur discovers new method of vaccination*

1898 *Martinus Beijerinck discovers viruses*

1928 *Alexander Fleming discovers penicillin mould*

1931 *Prototype electron microscope is invented*

1945 *Howard Florey and Ernst Chain receive Nobel Prize for world's first antibiotic*

1977 *Carl Woese discovers single-celled organisms called archea*

THE ROOTS OF MICROBIOLOGY

Before there was microbiology, there were many other ideas about where infectious diseases came from. In early civilizations, epidemics were thought to come as punishments from angry gods, like Nergal, the Mesopotamian god of pestilence. In Ancient Greece, people believed that the environment itself could cause disease – for instance, marshes or stagnant water were considered a breeding ground for infections. This notion eventually morphed into 'miasma theory', which held that infectious disease was spread through toxic air, contaminated by rotting organic matter; one only needed to breathe the putrid vapours in an unhygienic area, and infection would occur. The theory seemed logical – there is, after all, a correlation between unhygienic conditions and disease – but not because of the air. At any rate, miasma theory lasted well into the 19th century.

It was the writings of Girolamo Fracastoro, an Italian physician, that first popularized an opposing theory of contagious disease. In 1546, Fracastoro proposed that the root cause of epidemics was invisible seed-like spores that multiplied rapidly. These spores, Fracastoro argued, could cause infection by direct contact, through carriers such as clothing that had come into contact with the sick, or through the air. Frascatoro's theory was one of the first insights into how diseases are spread through invisible microbes, and it would go on to influence the early generations of microbiologists.

Epidemics were once thought of as punishments from angry gods. In this woodcut, the Christian God inflicts three days of pestilence.

Spontaneous Generation

In the 17th century, thinkers became preoccupied with the resurgence of an old debate about the origins of life. Originally conceived in Ancient Greece, the theory of spontaneous generation held that life could emerge even without the presence of other living matter. Under this theory – known as abiogenesis – it would be entirely possible for insects or mammals to just come into existence, rather than through being born of another living creature. Proponents pointed to the example of maggots growing in rotting meat; to them, it seemed as though the maggots came out of nowhere.

But plenty of scientists thought otherwise, and in 1668, an Italian physician named Francesco Redi demonstrated that

abiogenesis was impossible. To show this, he filled a series of jars with decomposing meat; he covered some of the jars with netting and left others entirely open. In the open jars, maggots grew inside the meat; in the jars covered with netting, flies were unable to enter to lay their eggs, and so no maggots appeared.

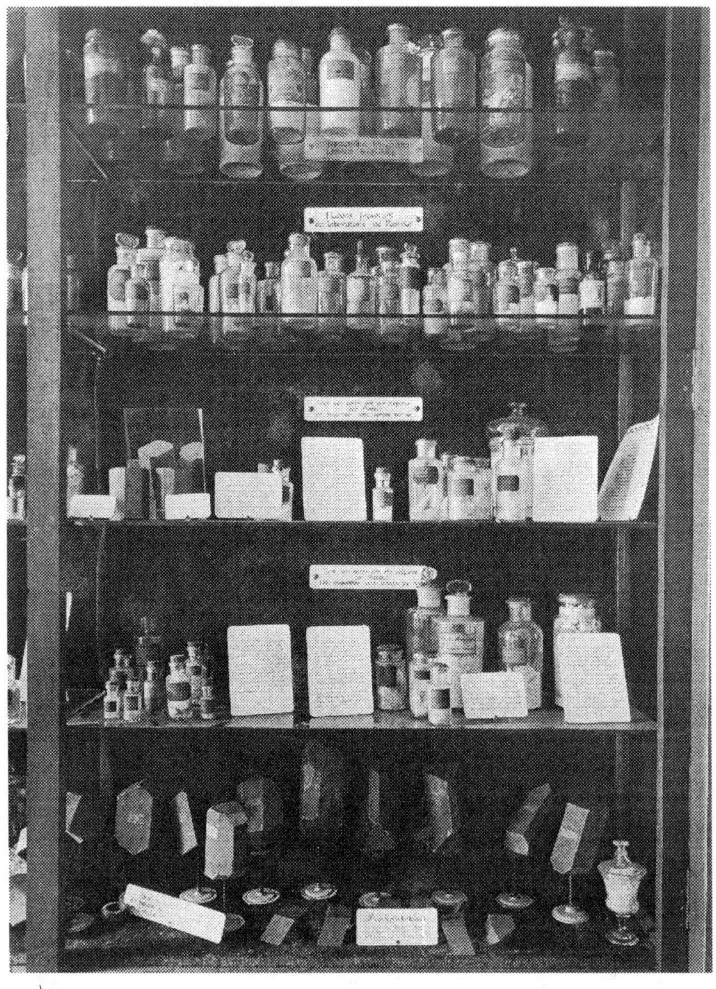

The flasks and other equipment used by Louis Pasteur in his laboratory in Paris.

The notion of abiogenesis was finally put to rest by French chemist Louis Pasteur in the mid-19th century. Pasteur was curious about where microorganisms came from, and he wasn't convinced by those who argued that microbes were the product of spontaneous generation. To test his theory, Pasteur set up an experiment involving a broth-filled beaker with a curved spout. Microorganisms in the air were unable to reach the broth through the bended spout, so no life grew in the broth. But when Pasteur removed the spout, the broth quickly became cloudy and decayed. Pasteur had demonstrated that microbes don't spontaneously generate within the broth, but that they enter into it from outside, through the air. Ultimately, Pasteur went on to demonstrate that microorganisms reproduce, just like all other living organisms, leaving the theory of spontaneous generation dead and gone.

The birth of microbiology as a discipline began with the study of a Dutch draper named Antonie van Leeuwenhoek. As a hobby, Leeuwenhoek ground lenses to create microscopes. Over some years, he perfected his technique, ultimately creating a microscope with many times the magnification of previous models. Through this microscope, Leeuwenhoek explored drops of rain and pond water, saliva, semen and stool, and was able to identify microorganisms – such as spermatozoa and infusoria – that no one had ever recorded before. He called these microscopic organisms 'animalcules'. In 1676, Leeuwenhoek saw a new type of animalcule through his microscope: a crescent-shaped microorganism originating in the human mouth. Although he didn't yet realize it, Leeuwenhoek had just discovered bacteria. But it took a while for the rest of the science community to catch up.

MICROBIOLOGY IN THE 19TH CENTURY

Over the course of the 19th century, it became clear just how widespread and powerful microbes are. In 1837, a German physiologist and a French engineer – Theodor Schwann and Charles Cagniard de La

The Dutch microsopy expert Antonie van Leeuwenhoek discovered bacteria.

Tour – both demonstrated that the process of fermentation is caused by microorganisms. Until now, scientists had assumed that fermentation was just a chemical reaction, but these findings demonstrated that yeast is actually alive.

Who's Who – Ignaz Semmelweis

Born in Hungary in 1818, Ignaz Semmelweis would go on to revolutionize medical practice through the introduction of antiseptic procedures, such as hand washing, which reduced microbial infection.

After training as a physician, Semmelweis was appointed to work at an obstetric clinic in Vienna, Austria, in the 1840s. Here, he noted a high level of mortality in new mothers following births attended by doctors and medical students rather than midwives. Scores of women at the hospital were dying from puerperal fever, an infection of the reproductive organs that can occur after birth. Semmelweis hypothesised that there was a connection between these deaths and the autopsies that the doctors and medical students also performed at work. He proposed that puerperal fever was a contagious disease that was passed to patients from the bodies of those who had died of the disease, via the hands of medical staff.

Semmelweis ran a trial in which clinicians washed their hands in chlorinated lime before examining patients; the mortality rate dropped from over 18 per cent to just over 1 per cent. However, when Semmelweis published his theories, he was met with scepticism. Many physicians believed that infections were unavoidable, and were the result of sheer chance.

Following this rejection, Semmelweis suffered a nervous breakdown and was admitted to a psychiatric hospital, where he died shortly afterwards from an infection. Although he didn't live to see it, Semmelweis' work inspired pioneers of antisepsis, such as the British surgeon Joseph Lister. The 'Semmelweis reflex' is now used to describe the innate reflex that humans have to reject new ideas and evidence when it contradicts established norms.

Throughout the 1850s and 1860s, the French chemist and microbiologist Louis Pasteur performed a series of experiments that opened up the science of microbiology. It seemed clear to Pasteur that if microorganisms could change the state of food, then they could have a role to play in health and disease as well. In 1865, Pasteur revealed a new theory about the plague of disease that had been decimating France's silkworm industry. He found that healthy silkworms generally became ill when they came into contact with those that were diseased, but that the spread of disease stopped if infected silkworms were isolated and the environment sterilized. Paired with his discovery of microbes in the air, this work helped Pasteur to establish the germ theory of disease, which holds that infections are caused by invisible microbes and can be passed between people through direct or indirect contact.

Did You Know?

Pasteur's work on microbes led to a number of new innovations to prevent food spoilage, including – of course – pasteurization!

But Pasteur's experiments with microbes and disease didn't end there. In 1879, he tried injecting a weakened version of a bacterium into healthy chickens. The chickens became temporarily ill, but they didn't die as he expected. When Pasteur then attempted injecting the same chickens with the deadly, full-strength version of the bacterium, he found that they did not become sick at all. Pasteur had discovered a new method of vaccination, and he quickly set about developing vaccines for other diseases, including anthrax and rabies.

The germ theory of disease became firmly embedded the following decade, when a German scientist named Robert Koch developed a method of demonstrating that specific microbes are responsible for specific diseases. 'Koch's postulates', as they became known, laid out a process that could be used to establish a causative relationship between a microbe and a disease. Koch used his method – which involved isolating microbes and testing their effects on healthy organisms – to identify the bacteria that cause anthrax, tuberculosis and cholera. The years that followed saw a surge in discoveries of bacteria and their links to specific diseases; scientists' knowledge of the microbial world was expanding.

In the late 19th century, microbiologists became aware of a new type of threat: viruses. In the 1890s, a Dutch microbiologist named Martinus Beijerinck was exploring a disease that affected tobacco plants. He attempted to isolate the bacterium behind the disease by filtering it out, as was the usual practice, but nothing was caught in the filter. This suggested that whatever was behind the disease was smaller than normal

Robert Koch worked out how to demonstrate that specific microbes are responsible for specific diseases.

bacteria. Beijerinck realized that this very tiny microbe was something entirely new – and he named his discovery a 'virus'. It quickly became clear that viruses were behind a number of other diseases, and that they were an entirely separate type of microbe.

By the turn of the century, the premises of microbiology had gone from theory to established reality. As our understanding of the microscopic world progressed, so too did our understanding of life itself – where it comes from and how it is defined. As the 20th century dawned, the foundations were set for the field to branch off into new areas, and over the course of the next 100 years, microbiology came to intersect with a vast range of different scientific fields.

THE ROAD TO MODERN MICROBIOLOGY

During the 20th century, microbiology flourished, expanding outwards into new areas of science. In the first decades of the century, vaccines had already been around for some time, providing protection from dangerous viruses, but there was no equivalent protection from bacterial infections. Thankfully, that began to change in 1928, when Alexander Fleming – a Scottish bacteriologist studying staphylococcus bacteria – returned from a holiday to find his bacteria destroyed. The samples had become contaminated by a mould, which seemed to have eaten away at the staphylococcus and prevented it from growing. Fleming recognized the potential of this mould as a medicine against infection, but he didn't have the resources or chemical knowledge to develop it into a drug. In the 1940s, a pair of better equipped scientists, Howard Florey and Ernst Chain, developed Fleming's mould into penicillin, the world's first antibiotic. For the first time, humans had a strong line of defence against bacteria, and the discovery has since gone on to save in the region of 200 million lives. But in his Nobel Prize acceptance speech, Alexander Fleming cautioned of the dangers of overuse of antibiotics, warning that they could become ineffective over time: an issue we're now facing today.

Taxonomy is concerned with the classification of things, usually living organisms, based on their shared characteristics.

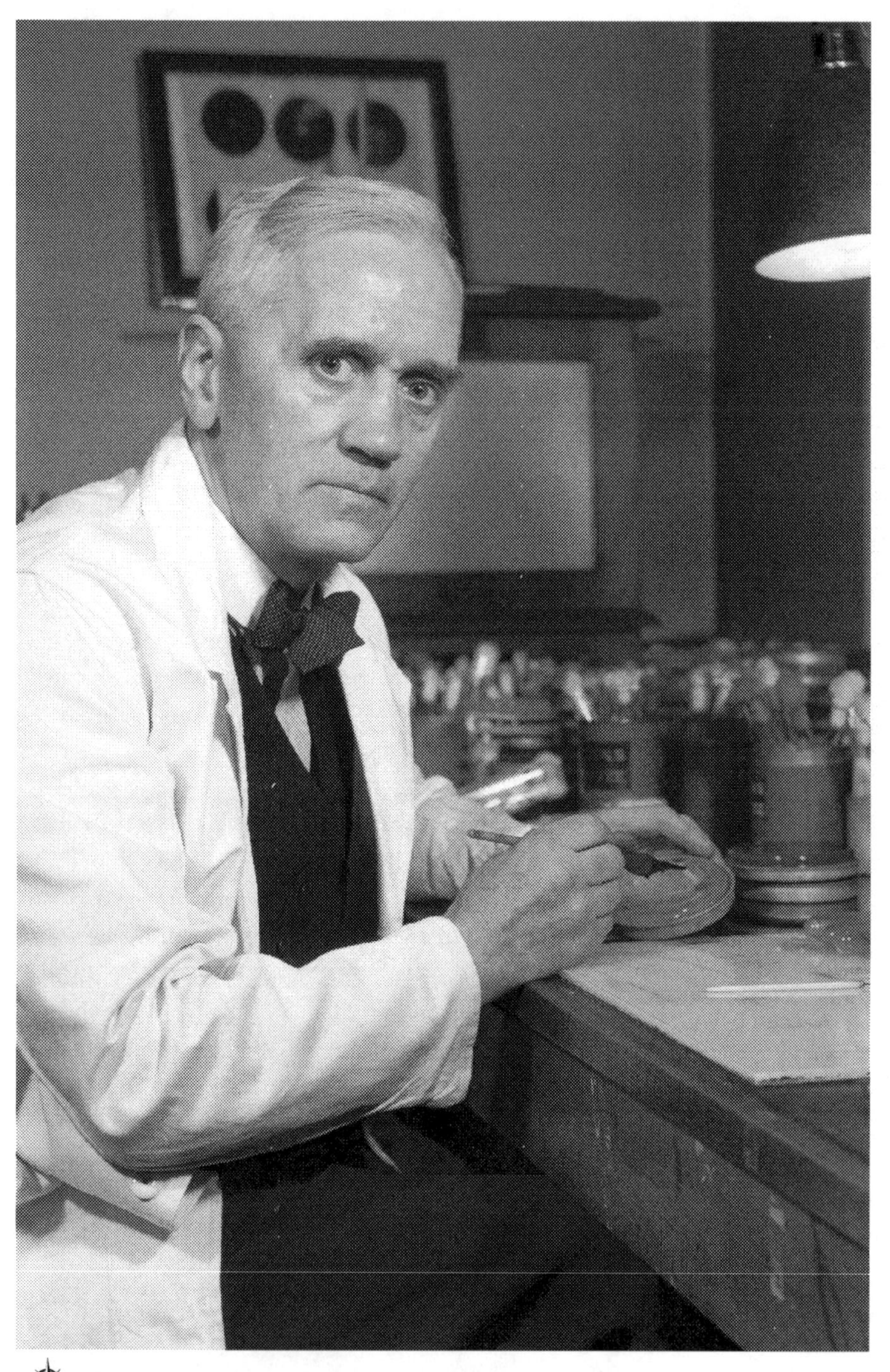

Alexander Fleming discovered pencillin, the basis for antibiotics, purely by accident.

In the same decade that Florey and Chain released their miracle drug, engineers developed the first electron microscope. This device used electrons rather than visible light to provide a more powerful image of a sample. With the help of electron microscopes, microbiologists became able to properly study viruses and other particularly small microorganisms. Meanwhile, the introduction of new cultivation methods meant that scientists could grow viruses in the laboratory and study them on demand, leading to rapid progress in microbiology. The ability to cultivate microbes effectively also supported the growth of biotechnology: a field in which living organisms are used to help manufacture commercial products, such as drugs. Using cultivated microorganisms, 20th-century scientists were able to produce medicines, including various vaccines and drugs, as well as industrial innovations such as pest-resistant crops.

Did You Know?

Viruses aren't really alive, at least not in any recognizable way. Unlike bacteria and other pathogens, viruses aren't made out of cells, so they can't independently reproduce. Instead, they need a host cell to provide them with the resources to grow.

From genetics to geochemistry, microbial research grew dramatically in scope during the 20th century. In structural biology, the dawn of DNA sequencing technology in the 1970s opened up entirely new avenues of investigation for microbiologists, who were then able to explore the genetic makeup of viruses, bacteria and other microorganisms. In the 1970s, an American microbiologist named Carl Woese introduced a new category of microbial life: archea – a group of single-celled organisms with unique characteristics. Archea are capable of surviving in some of Earth's most extreme environments, including boiling hot springs and the depths of the ocean, and they support a variety of different ecosystems.

Over the course of the 20th century, scientists uncovered the scale of microbial life and its importance to the planet, but there's still a great deal more to learn. Microorganisms are capable of many astonishing things, and learning how to harness those abilities is one of the next big challenges that microbiologists face. From improving agriculture to managing waste, microbes are already gaining pace in industry, and they have the potential to support all sorts of industrial processes.

But while microbial life can be useful to humankind, it can also be deadly. Nowadays we are much better at understanding how to exploit and respond to microorganisms, but it's important to remember what we're dealing with. As antibiotics become less effective and new epidemics begin to crop up around the world, we should be mindful of the power that microbes have over the environment around us, and the great lengths to which scientists have gone in order to protect us from these, Earth's most powerful lifeforms.

PART FIVE

THE 20TH AND 21ST CENTURIES

IN THE 19TH CENTURY, scientific advancements were often the products of pragmatic and practical-driven considerations. New machinery and medicines made life easier for everyday people and helped to strengthen nation states, and these were major drivers of scientific progress. In the 20th and 21st centuries, that focus has shifted to a more abstract and theoretical approach to research – an approach that has since transformed the scientific landscape.

The very first year of the 20th century saw the early outlines of a quantum theory of matter, followed closely by a revolution in accepted knowledge regarding the physical laws that govern nature. Over the years that followed, scientific horizons expanded to encompass invisible, almost imperceptible units of matter. At the same time, researchers began to actively explore the outer reaches of an inconceivable, infinite cosmos.

As has been the case throughout much of scientific history, these discoveries were inevitably spurred on and guided by geopolitical and social factors. Particle physics grew partly out of research into atomic weapons; astronomy and space science flourished as a result of Cold War competition. And, of course, the horrors of two world wars gave rise to a wealth of new technologies and ideas.

In turn, we've seen technological developments play a fundamental role in scientific progress. In the 20th century, the advent of advanced electron microscopes allowed scientists to explore the biological world in unimaginable detail; and powerful imaging and satellite technology enabled the launch of the Hubble Space Telescope in 1990, which has since provided us with an extraordinary perspective on far-flung reaches of the cosmos.

The 20th and 21st centuries have seen science unearth increasingly complex ideas and questions, many of which are yet to be fully resolved. At the dawn of the 21st century, the scientific landscape has been dramatically transformed. We can now look deeper, explore further and comprehend more than ever before, and what we've discovered is that the world is more complex than we could have ever imagined.

CHAPTER 13
PHYSICS

'One cannot help but be in awe when he contemplates the mysteries of eternity, of life, of the marvellous structure of reality.'

– ALBERT EINSTEIN (1879–1955), THEORETICAL PHYSICIST

It is sometimes said that physics is the purest form of science, focused as it is on the fundamental laws of nature. The study of energy, matter and the interaction between the two, physics is all about asking the big questions, such as: 'what is matter made of?' and 'what causes matter to behave in the way that it does?'

While the study of physics has a history that spans millennia, there's still a lot left to learn. The classical physics established by scientific giants such as Galileo Galilei and Isaac Newton looks very different from the modern physics of Albert Einstein, Max Planck and their contemporaries, but both are united by the same desire to understand the basic properties of nature. This is the story of how scientists have worked to unpick the mysteries of the universe, and how we've discovered that the physical world is more intricate than we ever imagined.

1638	*Galileo Galilei publishes theory of falling bodies*
1687	*Isaac Newton proposes three laws of motion and the law of gravity*
1820	*André-Marie Ampère theorizes existence of what we now call electrons*
1821	*Michael Faraday invents first electric motor*
1873	*James Clerk Maxwell publishes equations explaining electromagnetism*

1887 *Heinrich Hertz observes the photoelectric effect*

1905 *Albert Einstein proposes theory of special relativity*

1915 *Einstein follows up with general relativity*

1932 *Discovery of neutrons completes the atomic model*

2012 *Subatomic particle known as the Higgs Boson is discovered*

THE ROOTS OF PHYSICS

The early history of physics is intimately connected with the development of other scientific fields, such as philosophy, mathematics and astronomy. A recognizable form of the discipline began in Ancient Greece, where scholars were fascinated by the big questions that characterize physics, and worked to create theories of the universe and matter.

But the story really got going in late 16th-century Italy. Here, polymath Galileo Galilei set the foundations for classical mechanics – the physics of the motion of bodies – with his famous thought experiment. Galileo argued that if one discounts the air resistance that causes a feather to float serenely to the ground, all bodies fall at the same rate, whether they're heavy or light. This idea was important because it suggested that something other than weight is responsible for drawing objects towards the Earth.

In the 17th century, the British polymath Isaac Newton proposed an answer. In 1687, Newton published his three laws of motion and his law of gravity. Together, these theories explain why objects behave in the way that they do when in motion, when still and when force is applied to them. Newton proposed that gravity is a universal force which works to pull all objects in the universe closer together. Gravity, he argued, is responsible for something as small as drawing a falling apple to the ground, and as large as keeping the planets in orbit. Newton's work was groundbreaking, but there were gaps in the theory that scientists would return to in the 20th century.

 *Galileo's famous thought experiment involved dropping objects from the Leaning Tower of Pisa.*

Did You Know?

The concept of atoms has a long history. The Ancient Greeks were the first to come close to an accurate theory of matter, when in the 5th century BC, a Pre-Socratic philosopher named Democritus (see page 129) suggested that all matter is made up of infinite, indivisible particles called 'atoms'.

Thermodynamics

An important branch of physics known as thermodynamics had its start in the 18th century. Concerned with the relationship between heat and other forms of energy such as mechanical work (the amount of energy transferred by a force), thermodynamics came about largely because many physicists were interested in the science behind what causes objects to warm up and cool down.

In the 1780s, the French chemist Antoine Lavoisier (see also pages 126–9) proposed that heat was a type of fluid called 'caloric' that transferred from hot to cold objects – so an object's heat was dependent on how much caloric it had. Under caloric theory, heat is a 'conserved quantity', which means that it exists in a fixed amount; you can't create new heat from nothing or destroy it; it simply passes into another object. It's clear that there are a number of problems with this theory – it doesn't account for heat created by friction, for instance.

In 1798, American physicist Benjamin Thompson – known as Count Rumford – spent hours boring a hole through a cannon barrel that was submerged in water. Within 2–3 hours, the water began to boil as a result of the heat created by the friction, demonstrating that heat could be generated in indefinite amounts. Rumford's findings challenged caloric theory by showing that objects don't hold a fixed amount of fluid-like heat that can be transferred to other objects, because new heat can always be created.

Over the course of the 19th century, the science of thermodynamics began to take shape. In 1825, a French physicist named Nicolas Carnot worked out the science behind the processes that take place inside heat engines. Carnot's goal was to improve the efficiency of steam engines, but his work also helped to establish some of the basic tenets of how heat energy behaves. Over the following decades, scientists formulated the fundamental laws of thermodynamics, which hold that the amount of energy in the universe is constant – it can neither be destroyed nor created – and that the amount of that energy which is available for work decreases over time to a constant. In the years since, the laws of thermodynamics have become a guiding force in science, and are now a cornerstone of modern physics.

Throughout the 18th and 19th centuries, work on electricity ushered in an entirely new area of physics. In 1820, Danish chemist and physicist Hans Christian Ørsted discovered that when he ran an electrical current through a wire, it caused the magnetic needle on his compass to move position and follow the direction of the current. The same year, French physicist André-Marie Ampère demonstrated that two parallel wires could attract each other if the currents were flowing in the same direction, and repel each other if the currents were flowing in opposite directions. Ampère also found that the strength of this force was dependent on the length of the wires and the distance between them. He theorized that there was some sort of 'electrodynamic molecule' that carried the currents of both electricity and magnetism – Ampère was in fact describing what we now recognize as the electron, and laying the foundations for the science of electromagnetism.

From here, electromagnetism really took off. In 1821, a British physicist and chemist named Michael Faraday invented the first electric motor: a device which used the interaction between electric currents and magnetic fields to generate electromagnetic force and produce power. Ten years later, in 1831, Faraday showed that movement could be converted into electricity by using electromagnetic induction: a phenomenon that became the basis for generators such as wind turbines and hydroelectric plants.

Electromagnetic induction is the process by which a magnet is passed through a varying magnetic field to generate an electric current.

Michael Faraday invented the first electric motor.

In the second half of the 19th century, Scottish physicist James Clerk Maxwell synthesized and expanded on this and other research to articulate his revolutionary theory unifying electricity and magnetism. Maxwell developed a series of equations which explained the mathematics behind electromagnetism. He also formulated a unifying theory which held that electricity, magnetism and light were all part of the same phenomena – a wave-like energy that contains both magnetic and electric fields. Maxwell's work unravelled one of the greatest mysteries of classical physics, and set the stage for the new era of physics that was to follow.

MODERN PHYSICS

At the beginning of the 1900s, there was a sense that physics was almost finished. Everything that there was to discover about the universe had already been discovered, and our understanding of the laws of nature was virtually complete. But in the 20th century, a new generation of physicists

Created in 1831, Faraday's sparking coil was used to show that magnetism could produce a spark.

began to discover phenomena that existing laws could not account for. Before long, it became apparent that the classical physics established by Newton and his contemporaries explained only phenomena taking place in the observable universe. When physicists explored some of the behaviours of energy and matter at extremes – like at the smallest and fastest imaginable levels – they found that all the established rules were broken and new laws took over.

One of the earliest signs that something wasn't right with classical mechanics came in 1887, when a team of American physicists named Albert Michelson and Edward Morley performed what is sometimes called the most famous failed experiment in history. They wanted to prove the existence of aether: a mysterious substance thought to fill empty space in the universe. Physicists at the time believed that light travelled through aether in the same way that sound waves travel through air. Michelson and Morley attempted to demonstrate the existence of aether by examining the ways in which it interferes with the speed of light. But despite their best efforts, they found that the speed of light never changed.

This presented physicists with a puzzle, and in 1905, a German patent clerk named Albert Einstein proposed a solution: the theory of special relativity. Under Einstein's theory, there was no need for light to move through an aether or any other medium. He suggested that light remains constant no matter what, and that nothing can move more quickly than light. In fact, Einstein argued that the speed of light is the only constant there is – everything else, including time itself, is variable. Einstein suggested that time varies depending on where you are in space; so the nature of a minute, an hour or a year could vary depending on the speed at which an object is moving. For objects travelling very fast, Einstein reasoned, time physically slows down. Now, if this sounds a little confusing, that's because it is. Einstein's theory completely dismantled the conventional wisdom of the field, challenging the seeming logic and order of the universe.

But Einstein wasn't done yet. Shortly afterwards, in 1915, he presented his theory of general relativity, which took these ideas even further. This time, Einstein proposed that space and time are interlinked, and that because both are variable rather than fixed, they can be warped by large objects. A good analogy for this is the idea of a bowling ball in the centre of a trampoline – the bowling ball weighs down the fabric, causing any objects around the outside to fall to the centre. This, Einstein argued, was the phenomenon that we describe as gravity. Heavy objects, like the Earth or the Sun, warp the fabric of spacetime, pulling objects towards them. That's why planets orbit the Sun, and why apples and other objects are pulled towards the Earth. Once Einstein was finished, a new era of modern physics had been established in which nothing is constant except the speed of light, and observable phenomena, such as time and space, could not be trusted.

Did You Know?
Perhaps the most famous equation of all time, Einstein's $E=MC^2$ *describes the relation between energy and mass, demonstrating that they can be converted into one another.*

But it wasn't only at the speed of light that the laws of classical mechanics broke down; there were also problems when studying the scale of the very small. In 1897, a few short years before the dawn of the 20th century, a British physicist named J. J. Thomson demonstrated the existence of negatively charged particles flowing inside atoms; this new subatomic particle was later dubbed the 'electron'. Thomson had discovered the unit of matter responsible for carrying electric charge, but even more than that, he had shown that atoms – and, by extension, all matter in the universe – were divisible into even smaller constituent parts. In 1911, another British physicist, Ernest Rutherford, discovered the proton, a positively charged particle making up the nucleus of the atom. Then, in 1932, the discovery of neutrally charged neutrons, which accompany protons in the nucleus, completed the atomic model.

But there was still much to learn. In the early 19th century, British physicist Thomas Young discovered that when light was passed through two nearby slits, it behaved in unexpected ways. Rather than dispersing in a wave-like pattern as expected, the light spread as though it were made up of individual particles that interfered with one another. Years later, in 1887, German physicist Heinrich Hertz observed the 'photoelectric effect', through which some metals produce electrical currents when exposed to certain frequencies of light.

Who's Who – Arthur Stanley Eddington

Sir Arthur Stanley Eddington was a pioneer in the field of astrophysics, and he helped to establish some of the fundamental physical theories of the 20th century. Eddington proved himself to be one of the greatest minds of his generation, and his work spans physics, astronomy and philosophy.

One of the achievements Eddington is best remembered for is his proof of general relativity. When Albert Einstein published his theory in 1915, it took the science community some time to catch up. But one of the first scientists to embrace the theory was Eddington – who, by that time, had become well regarded within the British physics community and had risen to director of the prestigious Cambridge Observatory. Eddington's willingness to put his weight behind the theory helped to publicize and spread awareness of relativity more broadly.

During the 1919 solar eclipse, Eddington observed and successfully photographed stars whose light had been warped by the mass of the Sun. This phenomenon – known as 'gravitational lensing' – demonstrated Einstein's theory that the mass of heavy objects like the Sun could create a gravitational field, warping the light from the stars. Eddington's work helped to take general relativity from a curious notion to a widely established theory.

But Eddington's achievements extended well beyond relativity. Through his work in astrophysics, he helped to unpick the intricacies of various stellar processes, from determining the formula that explains the brightness of various stars to investigating their interior structure and characteristics.

Towards the end of his life, Eddington attempted to develop a unifying 'theory of everything', which could tie gravitation, relativity and quantum physics together in one neat package. Although he died before completing the work, Eddington's ideas have helped to inform the work of modern physicists who are still searching for an answer.

Eddington's contributions cannot be underestimated. As an astronomer, he helped to open up the field of stellar science; meanwhile, his work on relativity and physics allowed scholars to unravel some of the fundamental mysteries of the cosmos.

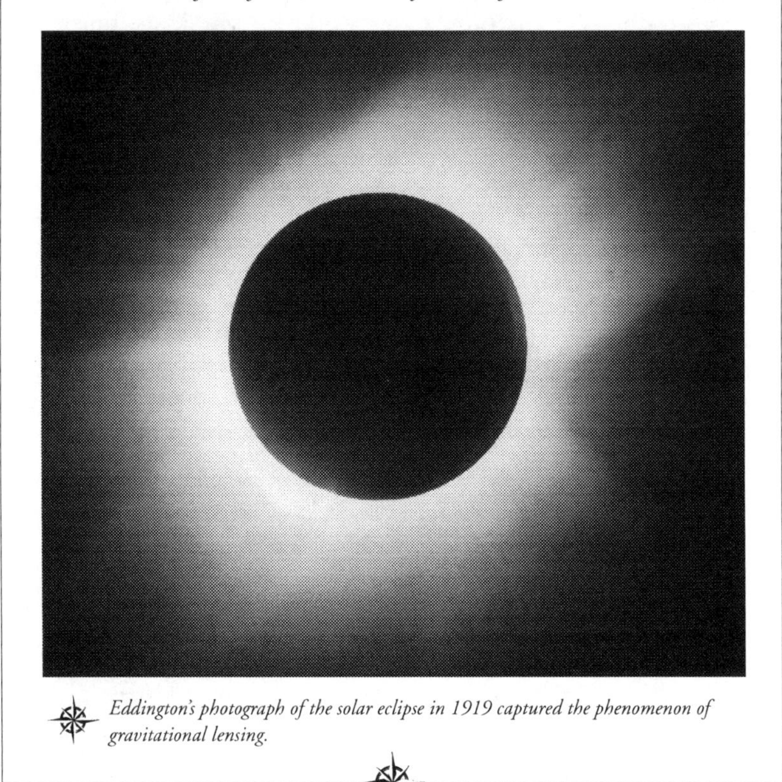

Eddington's photograph of the solar eclipse in 1919 captured the phenomenon of gravitational lensing.

In 1905, Einstein developed a theory that explained the phenomenon. He built upon an idea put forward some years earlier by another German physicist, Max Planck, who proposed that light might be 'quantised' – that is, that light travels in discrete packets, or particles, rather than in one continuous wave. The photoelectric effect, Einstein suggested, was caused by these particles of light, called 'photons', colliding with electrons on the surface of the metal. This interaction between the different particles was causing an electron to be ejected from the metal, and that ejected electron accounted for the current produced. Einstein's theory was revolutionary because it suggested that light was capable of behaving as both a wave and a particle – a seeming impossibility.

This was just the beginning for particle physics. In the mid-20th century, the discovery of fundamental particles such as quarks and fermions opened up an entirely new world for physicists, but it also revealed major holes in existing theories. As scientists began exploring the behaviour of these tiny particles, it became clear that the laws of classical mechanics did not seem to apply to them either.

Fundamental particles are smaller than subatomic particles such as electrons and neutrons, and are seemingly the smallest units of matter in the universe.

During the first half of the 20th century, physicists such as Werner Heisenberg and Erwin Schrödinger began to observe quantum phenomena: the unexpected behaviour of atomic and subatomic particles. Many of these behaviours are counterintuitive, and don't marry up with phenomena we observe on the larger scale. For instance, subatomic particles seem to be linked together and able to affect each other despite being separated by vast distances; and it's impossible to determine both the position and velocity of subatomic particles at the same time, because they change behaviour when observed. Quantum physics is still something of a mystery, and there's a lot left to learn about the laws of nature at the smallest scales. But what is clear is that the classical laws of physics are not sufficient to account for phenomena taking place on the quantum level.

With the benefit of hindsight, it's strange to think that, at the dawn of the 20th century, physicists felt that their work was almost done. We now know that the universe is so much more complex and unpredictable than previously thought, and that we're still a long way from understanding the many intricacies of nature. It's hard to predict what we'll uncover over the course of the next 100 years, but one thing is for sure: the work of physicists is nowhere near complete.

CHAPTER 14:
BRAIN SCIENCE

'The brain is the organ of destiny. It holds within its humming mechanism secrets that will determine the future of the human race.'

– WILDER PENFIELD (1891–1976), NEUROSURGEON

The human brain is the most complex object in the known universe. Inside your head is a computer more powerful than the most sophisticated technology that has ever been or will ever be created. There are more nerve cells (called neurons) in the brain than there are stars in the Milky Way. Each of these cells can be connected with thousands of others, to create over 100 trillion connections (known as synapses) in your brain. Together with the spinal cord – an information highway that connects your brain and body – the 3-pound organ in your head controls virtually every action you're aware of in your body. From processing your emotions to remembering to take out the trash, your brain is hard at work every day, keeping you in check.

There are broadly two separate ways of studying the magnificent organ inside our heads: neuroscience, which looks at the physical structure and function of the human brain; and psychology, which is the study of behaviour and the mind – our consciousness and the source of our thoughts, memories, decision-making, emotions, intellect, language and perception.

c. 5000 BC	*Neolithic societies practise trepanation as a pseudo-medical treatment*
c. 4th CENTURY BC	*Aristotle suggests the heart is the seat of consciousness*

c. 200 BC	*Galen suggests the brain is responsible for cognition*
1543	*Andreas Vesalius publishes anatomical sketch of the nervous system*
1811	*César Legallois demonstrates link between specific parts of the brain and their associated function*
1879	*Wilhelm Wundt founds first laboratory for experimental psychology*
1896	*Sigmund Freud coins the term 'psychoanalysis' for his new treatment*
1902	*Ivan Pavlov demonstrates conditioning through his studies on dogs*
1906	*Santiago Ramón y Cajal is awarded Nobel Prize for discovering the neuron doctrine*
1973	*First MRI scan image is published*

While both neuroscientists and psychologists consider more than just the brain in isolation, we can broadly group the two separate disciplines under the term 'brain science'. Over the centuries, scientists and philosophers alike have sought to answer questions about the nature of thought, perception and self. Foremost among these is the question of how our mind relates to our brain and nervous system, the so-called 'Mind-Body problem'.

The brain consists of white and grey matter. White matter contains axons and oligodendrocytes, while grey matter contains neurons.

These are questions not just about matter, but about ideas. How do we define 'normal'? Is there such a thing as 'normal'? What does it mean if there is no 'normal'? The science of the mind is thus made up of biology, chemistry, philosophy, sociology, neurology, anthropology and physiology. It is important, because our place in the world is determined by our minds and those of others around us. Ultimately, the study of brain science helps to pick away at that perennial question: what does it mean, fundamentally, to be human?

THE ROOTS OF BRAIN SCIENCE

While the significance of the brain hasn't been recognized throughout all of history – indeed, for many years scholars believed the heart to be the seat of consciousness – neurological activity, such as brain damage and mental illness, has always existed, and ancient civilizations had their own approaches to dealing with these issues. For instance, as early as the Neolithic Period, people were practising trepanation, a pseudo-medical treatment that involved drilling holes into the skull. It's thought that ancient people may have believed that creating a hole in the brain could help to treat headaches, neurological pain, epilepsy, mental illness and even demonic possession.

In Ancient Egypt and Greece, the importance of the brain was not widely recognized. Instead, both Greek and Egyptian thinkers believed that the heart was the 'thinking' organ – the seat of consciousness, thought and emotion. In the 4th century BC, the Greek philosopher Aristotle argued that the main purpose of the brain was to cool the

Trepanation, which involved drilling a hole in the patient's skull, inspired this painting, Removing the Stone of Madness, *attributed to Jan Sanders van Hemessen,* c. *1550.*

heart, which was the centre of human intelligence and vitality. While he was wrong about the body's most important organ, Aristotle did help to set up one of the most enduring questions in brain science: the nature/nurture debate. Aristotle believed that infants were born as a blank canvas and that experience moulded the person. But his teacher, Plato, disagreed. Plato argued that behaviour and characteristics were innate within a person – present from the moment they came into being. We still don't have a definitive answer to this conundrum, and the nature versus nurture debate continues to rage in the 21st century.

Did You Know?
While there is no certainty as to why the practice of trepanning was performed, it could be that our ancestors believed the procedure could release bad spirits or alleviate pain, and it may have been used in response to migraines, seizures or mental health issues.

The significance of the brain was better understood by the Greek physician Galen. In the 2nd century, Galen studied the anatomy of the brain by dissecting various animals. He suggested that the brain was responsible for cognition and also for conscious action such as movement. Galen even theorized on the importance of the spinal cord and cranial nerves for controlling action.

Brain science continued to advance in the Islamic world, where medics paid particular attention to mental health and the relationship between mind and body. The Persian polymath Abu Zayd al-Balkhi proposed a link between sickness of the 'psyche' and sickness of the body, suggesting that the former could lead to anger, anxiety and sadness. In the 11th century, the polymath Ibn Sina described a series of neuropsychiatric conditions, including what we would now recognize as mania, dementia, stroke, depression, insomnia and schizophrenia.

At the beginning of the 16th century in Europe, advances in anatomy helped our further understanding of the brain. In his sketches of the human body, the Belgian anatomist Andreas Vesalius produced a map of the nervous system, revealing poorly understood regions such as the corpus callosum. Anatomy also helped to inform the thinking of 17th-century French philosopher René Descartes, who introduced the idea of 'dualism'; this theory held that the mind was a non-physical entity, separate from the body, and that mind and body worked together to shape the human experience.

Now known as neurology, the branch of medicine that works with the brain and surrounding nervous system was founded in the late 17th century by British physician Thomas Willis. Through a series of texts, Willis published his in-depth studies of the nervous system and associated diseases, becoming the first person to describe the arteries that provide the brain with its blood supply and the spinal accessory nerve that powers the neck.

At the same time as the foundations of neuroscience and psychology were being laid, growing interest in the significance of the mind also helped to trigger some more spurious practices. In the 1770s, an Austrian physician named Franz Mesmer developed the practice of mesmerism, which held that the body acted according to the laws of magnetism, and that it could be regulated and healed through inducing trace-like states. Mesmerism became exceedingly fashionable in Europe for a time, until it was exposed in the late 17th century by a team of scientists assembled by the French King Louis XVI.

Phrenology, another pseudoscience, also became very popular around this time. Practitioners believed that the shape of skull could be used to study brain shape and thus diagnose mental abilities, dispositions and disorders. Phrenology became so popular in Europe and the US that some employers even demanded that phrenologists assess an applicant's suitability for a role.

The central nervous system, consisting of the brain and spinal cord, controls information and activity in all parts of the body. These organs are so vital that they are protected by the skull and spine, respectively.

Parisian patients receiving mesmeric therapy, a fashionable 18th-century pseudoscience.

In the early years of the 19th century, a French physician named César Julien Jean Legallois became the first to demonstrate the link between a particular part of the brain and its associated function. In 1811, Legallois revealed through experiments using rabbits that the medulla oblongata (part of the hindbrain) is responsible for respiration. From here, other scientists began to unravel the links between different areas of the brain and their functions, providing more sophisticated understanding of the various regions and their physiological roles.

Neuroscience had been established in its early form by the turn of the 20th century; at the same time, psychology was beginning to gain recognition as a scientific discipline, independent of philosophy. In Leipzig in 1879, a German physiologist named Wilhelm Wundt founded the first laboratory for experimental psychology. The father of modern psychology, Wundt used scientific research methods to study mental processes, establishing an empirical approach to study of the mind and creating the foundations for the modern era of brain science.

Did You Know?

In 1848, an American railway worker named Phineas Gage suffered a catastrophic injury when an iron rod pierced the left frontal lobe of his brain. Although Gage survived, his personality changed. The case was important because it established the role that the brain plays in shaping personality.

Mental Health

Our understanding of mental illness has been limited for much of human history. In the ancient world and for many years afterwards, psychological issues were often considered a punishment sent by the gods. However, there were some thinkers who challenged this view. The physician Hippocrates identified diseases of the mind as originating with the brain, and suggested that psychological issues – just like other diseases – were not supernatural in nature, but rather the result of physical causes.

In the medieval Islamic world, Persian physicians and scientists actively explored the concept of mental health. Between the 9th and 11th centuries, the polymaths Ibn al-Razi and later Ibn Sina studied the diagnosis and treatment of mental illness, documenting their observations on mental disorder and psychosomatic medicine. This thoughtful approach was reflected in mental health care. Thirteenth-century Islamic hospitals, called Bimaristans, included wards for mentally ill patients who were treated with music and bathing. Indeed, Al-Fustat, a hospital in Cairo, is thought to have been the first to offer treatment for mental disorders all the way back in the 9th century.

Medieval mental health treatments in Europe were generally less advanced, and treatments could often be counterproductive and profoundly cruel. Bethlem Royal Hospital in London was one of the first hospitals for mental health in Europe, operating from the late

The asylum at Bethlem, one of the first hospitals for mental health in Europe.

14th century, and was famed for the callousness of its approach to patients.

There was a challenge to this cruelty in the late 18th century, during the period of the enlightenment. A French physician named Philippe Pinel and his hospital superintendent Jean-Baptiste Pussin ordered restraints to be removed from patients at their hospital, in favour of a psychological approach that involved talking with patients and assessing their case history. In England, William Tuke, a Quaker, founded the York Retreat in response to the death of a fellow Quaker in the then appalling conditions at the York Asylum. Tuke focused on the humanity and dignity of mentally ill people, and their need to be occupied in positive tasks to promote recovery.

By the 20th century, mental health was becoming better understood as a physical condition tied to the brain. Psychological *therapies became well established, alongside pharmacological treatments such as selective serotonin reuptake inhibitors (SSRIs). While there is still much to learn about psychological conditions, we've come a long way in our understanding and approach to mental health.*

MODERN BRAIN SCIENCE

At the turn of the 20th century, Santiago Ramón y Cajal discovered, through his studies on brain tissue, that the nervous system is made up of separate, individual cells (called neurons) and gila rather than a continuous network – an achievement that would earn him a Nobel Prize in 1906. Cajal's theory was called the neuron doctrine, and it established the importance of neurons within the nervous system. Shortly after this, in 1932, another Nobel Prize was jointly awarded to two British physiologists, Charles Sherrington and Edgar Adrian, for discovering the importance of synapses. These tiny junctions between neurons are the channel through which electrical and chemical signals are passed, allowing messages to be communicated within the nervous system.

Did You Know?

There are more than 86 billion neurons – cells that transmit nerve impulses like movement or pain – in the human brain.

Meanwhile, psychology was undergoing a transformation of its own. At the turn of the 20th century, an Austrian physician and former neuroscientist named Sigmund Freud began popularizing a new kind of medical treatment: psychoanalysis. Freud had trained under the great neuroscientists of the day, including Cajal himself, but after becoming interested in hypnosis, his attentions shifted away from the physical brain and towards the idea of the unconscious mind.

Beginning in the 1890s, Freud began developing his own form of psychiatric therapy based on hypnosis, dream analysis and free interpretation, believing these to be paths to the unconscious mind. Freud theorized that the human mind was layered with unacknowledged beliefs and desires, many of them rooted in sexual development. While the scientific merit of Freud's psychoanalysis is widely questioned today, his work did help to popularize the notion of unconscious thought, and psychoanalysis undoubtedly had a lasting impact on Western psychiatry.

Freud's research partner for some years was the Swiss psychiatrist Carl Jung, who went on to leave his own indelible legacy on psychology. After parting ways with Freud in 1912, Jung proposed some of his most influential theories on personality and the mind. The notion of introverted and extroverted personalities originated with Jung, as did

Sigmund Freud (front row, left) and Carl Jung (front row, right) with colleagues, 1909.

the idea of individuation – the lifelong process of becoming one's true self by healing the split between the conscious and unconscious mind.

In the 1920s and 1930s, psychoanalysis fell out of favour, as psychologists sought to develop a more empirical and evidence-based approach to the science. Cue the rise of 'behaviourism', a school of thought that focused purely on observable, measurable behaviour. One of the earliest examples of behaviourism can be seen in the experiments of Russian physiologist Ivan Pavlov. In 1902, Pavlov demonstrated the principle of conditioned response – the way in which we learn through forming associations. Pavlov demonstrated this through his famous experiments with dogs. When a clicking metronome was introduced prior to feeding, over time Pavlov's dogs would learn to salivate at just the sound of the metronome, even in the absence of food – they had learned the association.

Who's Who – Stanley Milgram

The American psychologist Stanley Milgram is famous for his influential and highly contentious experiments on authority, obedience and conformity. Born in 1933 to a working-class Jewish family, Milgram grew up close to the tragedies of the Holocaust. Spurred by the desire to understand how such atrocities could be committed by human beings, Milgram carried out a series of experiments that went on to shape our understanding of social behaviour and psychology.

In 1961, shortly after earning his PhD in social psychology from Harvard, Milgram undertook his now famous shock experiment, which demonstrated the capacity of ordinary people to do terrible things in the name of obedience. During the experiment, participants were asked by a scientist in a white lab coat to administer increasingly high-voltage shocks to people whom they believed were other participants. The shocks were not real, and the people being shocked were actors, but the participants believed that they were in fact electrocuting real people. As the shocks appeared to grow increasingly more severe and dangerous, the majority of participants were willing to continue doing exactly what they were told to do, regardless of the consequences.

Milgram's experiment demonstrated how easy it was for human beings to carry out extreme actions if told to do so by a figure of authority – a finding that created reverberations throughout the psychological community. In 1971, another American psychologist, Philip Zimbardo, built upon Milgram's findings with his Stanford Prison Experiment. In this study, unknowing college students took the roles of guards and prisoners in a fake prison. Within six days, the experiment had to be stopped after the 'guards' became increasingly brutal towards their 'prisoners'. Although criticized both for its ethics and rigour, Zimbardo's experiment was influential in demonstrating the extent to which social pressures govern behaviour.

Stanley Milgram (right) and some of his colleagues pose with the 'shock box', a machine supposedly capable of delivering punishment shocks of between 30 and 450 volts.

Meanwhile, Milgram continued to produce influential studies throughout the remainder of his career. While the experiment that he performed could never be repeated today for clear ethical reasons, Milgram's work has undoubtedly helped to shape contemporary understanding of obedience and social behaviour.

In the 1950s and 1960s, a new approach began to take hold in psychology, emerging as the dominant movement that we now recognize today. Cognitive psychology rejected the behaviourists' narrow focus on observable behaviours; instead, they argued that mental processes such as memory, problem-solving and perception are vital to understanding human psychology. The growth of cognitive psychology was supported by a raft of new tools for studying the brain, including MRI and PET scans, which allowed researchers to visualize physiological activity taking place in the brain in relation to mental processes.

Cognitive psychology brought the discipline closer in line with the field of neuroscience. In the modern age, neuroscience and psychology are both recognized as distinct disciplines but, while there is still debate as to the extent of crossover, it's clear that there are links and that each discipline can support and inform the other. For instance, the field of neuropsychiatry treats mental disorders as diseases of the nervous system while also integrating environmental factors and mental processes of the patients.

In the 21st century, the science of the brain is reaching new heights of depth and complexity, yet there is still a great deal that we don't understand. In 2013, the US government launched a public-private research initiative aimed at improving our knowledge of brain function on the cellular level, with a view to improving treatments for disorders such as depression, Alzheimer's and Parkinson's. A number of other countries have since followed suit, leading to a surge in brain science research.

It's strange to think that the most sophisticated machine known to humankind is stored within our own skulls. We are still working to understand how our own minds function, and we are likely to have a long way to go before full understanding is ever achieved. But one thing is for sure: the only thing smart enough to unravel the mysteries of the human brain is, of course, the human brain.

CHAPTER 15
COMPUTER SCIENCE

'The best way to predict the future is to invent it.'

– ALAN KAY (b. 1940), COMPUTER SCIENTIST

It is sometimes said that, in a large city, you're never more than a few metres away from a rat; but in the digital age, it is probably more likely that you're never more than a few metres from a computer. Computer technology has become ubiquitous, and we're not just talking about laptops and PCs – alarms, watches, cars, phones, televisions, kitchen goods; computing technology underpins many different aspects of our lives.

Computer science is the study of computational systems, and it has become one of the most important and impactful sciences of the modern age. It is often said that we are currently in the midst of a digital revolution, a historical epoch that – like the Industrial Revolution that went before it – has transformed the everyday lives of people around the world.

But while the digital revolution is a profoundly modern phenomenon, it is also the product of a long legacy of computer science. Beginning with the mathematicians of ancient Mesopotamia, the foundations of computing stretch back many centuries. From the cogs and gears of the first mechanical calculators to the billions of transistors that power modern smartphones and tablets, computer technology is a very visible testament to the pace and power of scientific progress.

c. 1100 BC	*Abacus is used for computations in ancient Mesopotamia*
c. 200–100 BC	*Early use of protocomputers, such as the Antikythera mechanism*
c. 820	*Al-Khwarizmi establishes founding principles of algebra*

1703	*Gottfried Wilhelm Leibniz develops advanced binary number system*
1843	*Ada Lovelace creates the world's first computer programme*
1941	*Alan Turing deciphers the Enigma code at Bletchley Park*
1969	*US Department of Defense funds early iteration of the Internet*
1975	*First commercially available personal computers come on to the market*
1989	*Tim Berners-Lee invents the World Wide Web*
1997	*First social network site is established, triggering growth of Web 2.0*

THE ROOTS OF COMPUTER SCIENCE

In its purest sense, the roots of computer science stretch right the way back to ancient Mesopotamia, with the development of early mathematics. From at least 1100 BC, the abacus was in use in this region, constituting the earliest known computing device. Evidence suggests that the earliest protocomputers were developed from around 100 BC. In 1901, a device was found on an ancient shipwreck in the Mediterranean Sea, and the chunk of rusted metal would turn out to be one of the most momentous archaeological discoveries in the history of computer science. Known as the Antikythera mechanism, after the region in which it was discovered, the mechanism is now recognized as the world's earliest known analogue computer. Functioning like a mechanical calculator, the Antikythera mechanism used gears to calculate information about astronomical events, such as eclipses and the motions of the Sun. The knowledge required to create such a device clearly existed in the ancient world, but it appears to have been lost during the centuries that followed.

Computer technology began to advance again in the Islamic world during the medieval era. Astronomers, such as the Iranian polymath

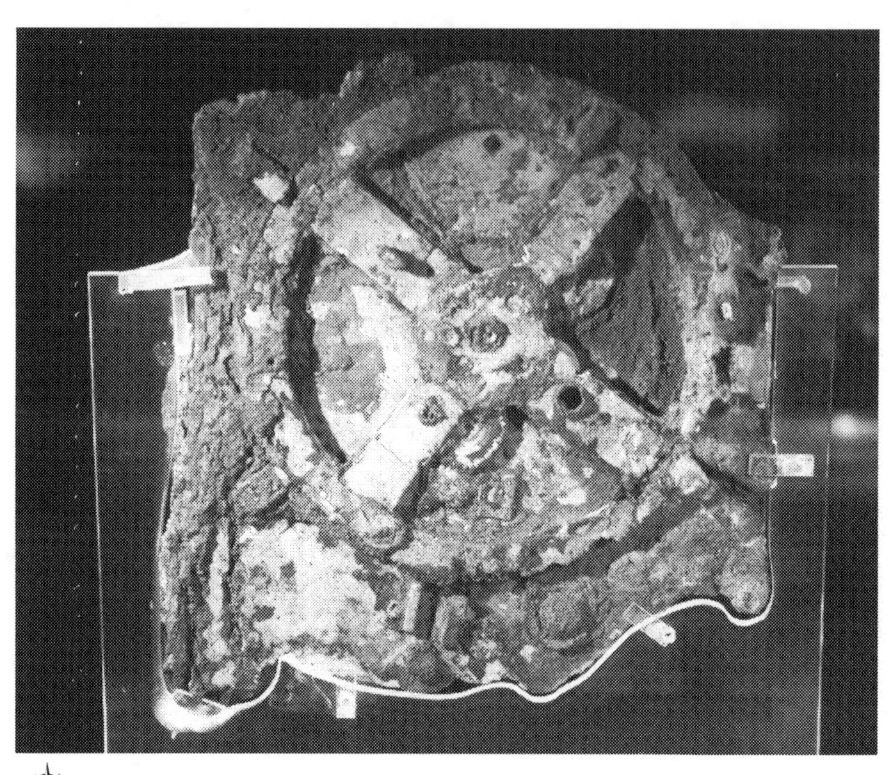

The Antikythera mechanism, an Ancient Greek computing device, created c. *100 BC.*

Al-Biruni, used an early form of mechanical computer to chart astronomical phenomena. Their devices, known as astrolabes, date back to the Hellenistic era, but they became much more sophisticated in the Islamic world. Islamic scholars also helped to develop some of the fundamental mathematical principles that now underpin computer science. In particular, the Persian mathematician Al-Khwarizmi (see pages 17–19) developed the science of algebra, laying the foundations for the systematic mathematics used in computer technology.

The earliest documented use of the term 'computer' was in 1613, although it referred not to a machine but to a person whose job it was to carry out calculations using a mechanical calculator. The first person recognized as a genuine computer scientist was the 17th-century German polymath Gottfried Wilhelm Leibniz. Leibniz helped to perfect the binary number system, which would later become the bedrock of modern computing. He also designed a more sophisticated model of the mechanical calculator, and was an early pioneer in the science of calculus. In addition to his scientific achievements, Leibniz was one of the first thinkers to propose a more expansive vision of what logical calculation could achieve, foreshadowing the growth of computer science that would take place in the centuries to come.

The first significant step towards something resembling a modern computer came in 1837, when English mathematician and engineer Charles Babbage proposed a machine called the Analytical Engine: a general-purpose computing machine able to carry out a variety of complex operations in a sequence, just as modern computer processors do. Babbage's colleague, the British mathematician and writer Ada Lovelace, designed a revolutionary algorithm to instruct the machine to perform calculations. But Lovelace was not only the world's first computer programmer, she was also a visionary who recognized that Babbage's computer could be used for much more complex activities, such as composing music and creating images. Lovelace understood that numbers could be used to represent different kinds of data, not just quantities, and that a machine capable of using these data would be able to do a whole lot more than just crunch numbers. Although Babbage and Lovelace never built the Analytical Engine, the concept of a general-purpose computing machine helped to expand ideas about what computers could do, and went on to inspire the modern era of computer science.

MODERN COMPUTER SCIENCE

The 20th century was the true dawn of the computer age. As with many technologies throughout history, computing developed partly as a result of the pressures of conflict and war. In the first half of the 20th century, the need for computing power to assist in codebreaking encouraged governments to invest in large-scale computing projects. Throughout the Second World War, both Allied and Axis forces used encrypted messages to keep their communications hidden from the other side. At centres like Bletchley Park in the UK and the Cipher Bureau in Poland, codebreakers such as the British mathematician Alan Turing made use of mechanical machines to decipher coded messages.

Ada Lovelace was a visionary who developed the world's first computer programme.

Did You Know?

Devices such as the Enigma machine were used to turn normal text into a code which could only be deciphered by someone with a matching machine and the correct settings, both of which were closely guarded secrets. However, through a combination of leaked documents, operational mistakes and computer-assisted deciphering techniques, Allied forces were able to break several versions of the Enigma code.

Who's Who – Grace Hopper

Grace Hopper was an American computer scientist and a rear admiral in the US Navy. Born in New York in 1906, Hopper went on to earn a PhD in mathematics at a time when very few women were able to attend university. In the midst of the Second World War, she joined the US Navy Reserves and, thanks to her background in mathematics, was assigned to work on an early computing project at Harvard University. She worked on the Harvard Mark I, an early computer used by the navy to aid the war effort. Hopper was among the first programmers of the Mark I, and developed a programme to help calculate the accurate firing of weapons. While working on the Harvard Mark I, Hopper coined the term 'computer bug', having discovered several physical bugs inside the machine, which kept interfering with the mechanical switches and causing faults.

In 1949, Hopper left Harvard for the Eckert-Mauchly Computer Corporation, and began work there on the UNIVAC I, one of the first commercial computers. In the years that followed, Hopper developed one of the world's first compilers: a programme that translated a programmer's written instructions into computer code. Hopper's compiler allowed programmers to write their instructions in a text-based language called COBOL rather than in complex machine code, making the process of programming machines much quicker and easier.

All the while, Hopper was rising through the ranks in the US Navy, and in the 1980s, she became one of just a handful of women appointed to the position of rear admiral. Hopper was 79 years old when she retired in 1986, making her one of the oldest officers on active service. Although she died in 1992, Hopper was posthumously honoured with the Presidential Medal of Freedom, presented by President Barack Obama in 2012, in acknowledgment of her remarkable contributions to computer science.

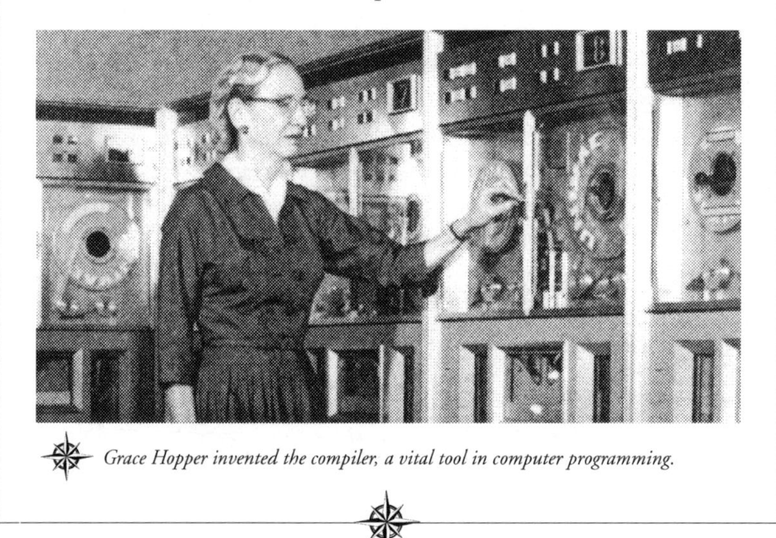

Grace Hopper invented the compiler, a vital tool in computer programming.

These early computers were huge and could take up entire rooms, and unlike modern machines, they used large, inefficient vacuum tubes: glass devices that controlled the flow of electric current. Electrical signals could be used to represent 'binary code' – a way of communicating information to computers that was based on only two quantities: ones and zeros. With these varying electrical signals, vacuum tubes could instruct a computer to perform specific actions using binary code. It was this vacuum tube technology that powered one of the world's first programmable general-purpose electronic computers, the ENIAC

(Electronic Numerical Integrator and Computer), which was built for the US Army in the Second World War. While they were revolutionary in scope, the first generation of modern computers were cumbersome, expensive and difficult to use.

Technicians connecting wires on ENIAC, one of the first programmable general-purpose electronic digital computers. Completed in 1945, ENIAC was intially designed to calculate artillery firing tables.

Binary

Digital information is represented in a series of ones and zeros in a system called 'binary', and this is the fundamental language of modern computing. Binary is a language built on just two different states: 'true' and 'false', or 1 and 0.

Binary is a useful system for computing because it allows for the use of Boolean algebra (see page 19), a mathematical system which uses true or false statements to build logic arguments. In computing, a zero can be thought of as false and a one as true. Using this simple system, computer scientists can build large, complex systems to carry out logical and arithmetic operations.

Binary is communicated through electrical signals, and over the decades, scientists have developed lots of different ways of producing these signals. In the early years of computing, binary was expressed through physical mechanical switches, which could signify either one or zero based on whether they were turned on or off. These switches could move between on and off many times a second, allowing for quick calculations compared to anything a human could achieve.

As time went on, new means of controlling electrical signals were developed, allowing for much faster binary communication. Modern transistors can now switch on and off billions of times every second; but even the most state-of-the-art computers still rely on the simple binary structure of ones and zeros to perform every function.

In the 1950s, a new technology took over from vacuum tubes as the main way of performing computer processes, leading to the emergence of second-generation computers. Transistors were much smaller than vacuum tubes, were made of plastic rather than glass, and could switch from on to off tens of thousands of times a second. Shortly afterwards, in the 1960s, the third generation of computer technology was marked by the invention of the integrated circuit, also known as the microchip. These tiny silicon chips were able to hold multiple transistors and other circuitry all in one place, making computers smaller and more efficient than ever before. Third-generation computers were more recognizable as the devices we know today, and users could interact with them using keyboards, monitors and a central operating system. For the first time, computers were becoming a commercially viable product for a mass audience, but it would still be some years before computers were affordable and efficient enough to become the ubiquitous machines we recognize today.

A computer programme is a set of instructions that a computer can understand and execute. Programming is the foundation for modern computing technology.

The fourth, and so far, final generation of computers began with the introduction of the first commercial microprocessor in 1971. Where previous models of computers had contained many different microchips, fourth-generation computers brought together most of the important computing functions in a single chip. This central processing unit, or CPU, meant that computers could be made even smaller and more

The Apple II was one of the earliest personal computers, first released in 1977.

powerful than before. Thanks to the rapid advances in processing technology and the development of more user-friendly interfaces, computer technology began to move out of research laboratories and into people's homes. In 1977, three different varieties of personal computer were released to the public, and by the 1980s, computers were becoming more and more common.

Did You Know?
In 1965, an early pioneer of computer chip development named Gordon Moore suggested that the number of transistors that you could fit into a computer chip would double every two years. This prediction has so far proved to be true, and is now known as Moore's Law.

Although computers are an integral part of modern life, taken alone they only tell part of the story of the digital age. In 1969, the US Department of Defense funded a project which aimed to allow multiple computers to communicate across large distances over a network. The project was entitled the Advanced Research Projects Agency Network, or ARPANET, and it laid the foundation for what would eventually become known as the Internet. Similar networks began to spring up for academics and researchers to share data and results, and soon the networks grew large enough that Internet services became useful to commercial customers, and the task of managing the networks was passed over to Internet Service Providers (ISPs). The Internet we know today can be thought of as a super-network that any computer can connect to.

A network is a group of computers, sometimes called servers, linked together. The Internet is essentially a network of networks.

But the most revolutionary development was yet to come. Today, the term 'web' is used synonymously with the Internet, but the web was actually a separate invention developed in the late 1980s, at a time when the Internet was already over a decade old. Designed by British engineer and computer scientist Tim Berners-Lee, the World Wide Web is a way of accessing information on the Internet. The web uses a set of rules called HyperText Transfer Protocol (HTTP) and web browsers such as Firefox, Chrome and Internet Explorer to access documents stored on the Internet; these documents are also known as webpages.

In the early 2000s, the phrase 'Web 2.0' became the newest buzzword; it referred to websites which, instead of containing static information that users could passively access, contained dynamic community spaces and user-submitted content. Social networks are a key part of Web 2.0, and today, sites such as Facebook, Reddit and Twitter are among the most popular sites on the web.

The history of computers has largely been an exercise in making ever smaller, ever faster devices, and this trend continues today with devices that contain billions of transistors so small you would need a microscope to see them. With so much power in such a small space, computers have found applications in ever smaller devices, from smartphones to watches. The miniaturization of computers has also led to more embedded systems, in which a computer is built into another device for the sole purpose of interacting with or controlling that device. Unlike general-purpose computers, embedded systems do not usually have applications beyond their primary function, so you might not immediately notice them – but they are everywhere, from your car to your alarm clock. Increasingly, these devices are being equipped with Internet capability, allowing them to communicate with both users and other devices, and together, this network of 'smart' devices is referred to as the Internet of Things (IoT). IoT allows us to control many devices usually through just one machine – so a user can change the temperature, record a film or turn on the lights with a single control device, such as a smartphone or tablet.

Today, computer scientists continue to push at the boundaries of what computer technology can achieve. We are living in the midst of a digital revolution that shows no sign of slowing down. And just as someone born in the wake of the Industrial Revolution would find it difficult to imagine a world without machines and steam power, people born today would be hard-pressed to envision a world without computers. Computers are now inextricably interwoven with our everyday lives, impacting on every aspect of contemporary culture; truly, computer science has reshaped the landscape of modern life.

CHAPTER 16
GENETICS

'Our own genomes carry the story of evolution, written in DNA, the language of molecular genetics, and the narrative is unmistakable.'

– KENNETH R. MILLER (b. 1948), CELL BIOLOGIST AND MOLECULAR BIOLOGIST

Genetics is the study of genes and heredity, and involves observing how genes work, how they can vary and mutate, how they interact with the environment and the role they play in biological processes such as health, disease and ageing.

Our genes are what make us us. Each one of us has approximately 24,000 of these small sections of DNA inside of us. They act like a blueprint, and their primary purpose is to instruct the body to create different kinds of proteins, each of which has an important function in supporting life. In this way, our genes control everything from the structure and function of organs to the colour of eyes and hair.

So where do genes come from? Well, we inherit them from our parents, who inherit them from their parents, and so on. Each parent passes down half of their offspring's genetic code through the sperm and the egg – that makes genes the fundamental units responsible for heredity; they're the reason you look like the rest of your family.

Working out the intricacies of these elaborate biological processes has taken many years and involved many great minds, and even now, there's still much more left to learn.

c. 4000 BC *Babylonian tablet documenting the pedigree of horses is created*

c. 400 BC *Hippocrates proposes early theory of inheritance*

1856	*Gregor Mendel begins revolutionary studies of pea plant genetics*
1868	*Charles Darwin sets out theory of inheritance by pangenesis*
1910	*Thomas Hunt Morgan shows that chromosomes are vital for genetics*
1952	*Rosalind Franklin creates Photo 51, a groundbreaking image of DNA*
1953	*James Watson and Francis Crick publish discovery of the double helix*
1978	*Nobel Prize is awarded for discovery of restriction enzymes vital to genetic engineering*
1990	*Gene therapy successfully trialled for the first time*
2003	*Completion of the Human Genome Project*

THE ROOTS OF GENETICS

It's tempting to think of genetics as a fairly new science, because we simply haven't known about genes for all that long, but its roots stretch back thousands of years. Since the earliest days of civilization, humanity has looked to understand and exploit genetic inheritance. Plant and animal breeding began in the Neolithic Era, as farming communities sought to create the strongest and healthiest varieties of crops and livestock. A Babylonian tablet dating back 6,000 years listed the pedigrees of horses alongside inherited characteristics such as height, and similar records exist for the breeding of plants. Indeed, for tens of thousands of years, humans have bred living things to become domesticated, physically altering their DNA in the process. It only takes one glance at the differences between a wolf and a chihuahua to recognize the deep genetic divergences that humans have engendered in the animal kingdom. But while early societies may have recognized the significance of genetics, they had no understanding of the mechanism.

Proteins are complex molecules which are the building blocks of life. They control the structure and operation of an organism's tissues and organs.

In Ancient Greece, the great physician Hippocrates proposed that the characteristics of parents were passed on through invisible material; this material, he suggested, originated from various organs in the parents and was reassembled in the mother's womb to form another person. He also suggested that acquired characteristics – that is, features that individuals acquire during their lifetime, such as big muscles or mutilations – could also be passed from parent to child. Of course, there were clear holes in this theory, for instance, why are the children of amputees born with limbs intact?

Despite these flaws, Hippocrates' ideas carried on influencing scholars through the centuries. Echoes of his theory were particularly prominent in the work of the British naturalist Charles Darwin, who is best known for his theory of natural selection: the idea that creatures evolve based on the random emergence of advantageous traits. Darwin understood that mutations which benefit an organism tend to be passed on through generations until they become the norm, but he didn't understand the process by which these traits were passed on. Building on the Hippocratic theory, he suggested that inheritance could be rooted in a process called 'pangenesis'. Darwin's pangenesis model held that cells inside organisms shed seed-like particles which gather in the reproductive organs. During conception, the particles from each parent are combined to form a new human, blended from the various seeds. Under this model, the entire organism is involved in passing on characteristics – the heart, the brain, the liver; every bit of tissue is involved in creating seeds from which to create a copy of itself. Darwin also suggested, as Hippocrates had before him, that acquired changes to the body of a parent could be passed on to that parent's children.

Evolutionary Biology

Biological evolution is the theory that groups of organisms change with the passage of time, causing their descendants to differ morphologically and physiologically from their ancestors. This process is dependent on the need of organisms to compete for the things that are essential to them individually to survive and reproduce. It is a continuing process that has been going on for millions of years rather than a finished event, although there is some debate over the degree to which human beings are evolving now.

Charles Darwin is generally credited with developing the theory of evolution but in reality, his was just one piece of work in a continuum that stretched as far back as the Ancient Greeks and well into later centuries. The Pre-Socratic Greek philosopher Anaximander proposed that animals could be transformed from one kind to another, while another Pre-Socratic thinker, Empedocles, speculated that creatures could be made up of various combinations of pre-existing parts. In 1377, the Arab historian Ibn Khaldun asserted that humans developed from 'the world of the monkeys' in a process by which 'species become more numerous'.

By the mid-19th century, ideas about evolution were already circulating when Darwin was formulating his revolutionary theory. Darwin's contribution was not to propose evolution, but to propose the mechanism of evolution; that is, natural selection. But even then, another British naturalist was hot on his tail. Alfred Russel Wallace was carrying out field research when the idea came to him that species evolved in accordance with their environment. Traits which gave individual animals within a species an advantage were passed down to the next generation, and the next, until those traits became commonplace. Wallace immediately wrote to his mentor in London, Charles Darwin, telling him of his revelation. Darwin, who had already arrived at a similar conclusion, invited Wallace to jointly author a paper with him, which the pair published in 1858. The following year, Darwin published On the Origin of Species, *and the theory of evolution by natural selection took off.*

In the years to come, scientists would bring together the disciplines of genetics and evolutionary biology, identifying the genetic mechanisms by which species evolve. In 1937, almost 70 years after Origin, the Ukrainian-American geneticist and evolutionary biologist Theodosius Dobzhansky produced Genetics and the Origin of Species. *In this text, Dobzhansky drew together the work of Mendel and Darwin, creating a modern synthesis between the two theories.*

By 1950, acceptance of Darwin's theory of evolution by natural selection was universal among biologists, and it is now widely accepted in society and taught within schools. It would be wrong, however, to claim that evolutionary biology is a completed science. There is still much more to learn, particularly with regards to the role of genetics in evolution. For now, at least, evolutionary biology and genetics will continue to prompt intriguing scientific and philosophical questions.

Charles Darwin developed the theory of natural selection which explained how evolution occurred.

Soon after sharing his theory with the world, Darwin was challenged by his cousin, the English polymath Charles Galton. Galton injected blood from rabbits that were one colour into rabbits of a different colour. If Darwin were correct, this process ought to result in changes to the colour of the rabbits' offspring, but no such changes took place – the offspring looked just like their parents. This finding suggested that acquired characteristics could not be passed down, and counteracted the idea that every part of the organism was involved in heredity.

Did You Know?

Darwin believed that genetic mutations resulted from gemmules – the theorized 'seeds' of cells – arranged in the wrong order.

But even as all of this was going on, the true unit of heredity was being uncovered by an Austrian monk named Gregor Mendel. Between 1856 and 1865, Mendel spent his days patiently cultivating pea plants in the garden of the abbey in Brünn (now Brno in the Czech Republic). By crossing different varieties, he was able to observe the ways in which traits were or were not passed down through the generations. Mendel believed that inherited characteristics were passed down in single hereditary units, and he was right; we now know that unit as the gene. To demonstrate this, Mendel studied the transmission of traits such as height and colour between different generations of pea plant, and he used these observations to develop fundamental laws of inheritance.

Mendel recognized that inheritance was controlled by distinct genes responsible for specific traits; he also understood that there are different versions of every gene inside an organism – one inherited from the mother, and one from the father. When two parents have a child, they each pass on one version of each of their genes. This means that the child has a totally random assortment of different genetic combinations, which is why siblings don't look identical. Finally, Mendel observed an order to which traits were and were not passed on; some genes, he realized, were dominant and others were recessive. In 1865, Mendel released his results to the world, but his work was overlooked by the scientific community and he died in obscurity; it was not until decades later that Mendel's remarkable contributions to genetics were finally recognized.

Gregor Mendel (back row, second from right) established the theory of genetic inheritance.

Did You Know?

To test his genetic theory, Gregor Mendel spent almost a decade growing and studying around 28,000 different pea plants.

MODERN GENETICS

Had Darwin and his contemporaries been aware of Mendel's findings in the 1860s, it's likely that the field of genetics would have emerged many years earlier. As it happened, though, that crucial turning point came instead at the dawn of the 20th century, when researchers suddenly rediscovered Mendel's work and sprang into action. In 1902, an American geneticist named Walter Sutton and a German biologist named Theodor

Boveri independently discovered that genes were located on chromosomes: thread-like structures that were found in the cell nucleus and consisted of tightly coiled DNA. Sutton and Boveri theorized that chromosomes were the physical manifestation of Mendel's theoretical gene, but it would be some years before experimentation supported the theory.

In 1910, American geneticist Thomas Hunt Morgan published the results of his research into fruit fly genetics, confirming the Boveri-Sutton theory and demonstrating the significance of chromosomes for inheritance. The previous year, Morgan had noted a mutation in one of his male fruit flies, which caused it to have white rather than the usual red eyes. He bred the mutant fly and observed the ways in which the recessive white-eyed trait emerged through the following generations. Morgan found that only male flies ever inherited the white-eyed gene; he had discovered sex-linked inheritance, in which certain traits could only be passed on to one of the two sexes – for example, male pattern baldness. Sex variations in inherited traits are a result of genes being stored on only the X or the Y chromosome. Women have two X chromosomes, while men have an X and a Y chromosome, so any gene stored on the Y chromosome – for instance, the white-eyed fly gene – could only ever be inherited by males. Morgan's student, Alfred Sturtevant, went on to create the first ever genetic map, charting the location of various specific genes on the fruit fly chromosome. In the 1940s, another American geneticist, Barbara McClintock, discovered that genes' position on the chromosome can change, and that genes are capable of 'jumping' to different locations, creating greater genetic diversity and sometimes triggering mutation and evolution.

Now that chromosomes had been established as the carriers of hereditary information, attention turned to the specific molecule that contains our genetic code: DNA (deoxyribonucleic acid). In 1952, the British chemist Rosalind Franklin and her student Raymond Gosling created one of the most important images in scientific history: Photo 51. The image provided a vital clue as to the molecular structure of DNA, and the following year, the American-British duo James Watson and Francis Crick determined the shape of DNA: the famous double helix. Their model demonstrated that the double helical structure, made up of two enjoined strands of genetic information, allowed DNA to separate and create copies of itself. Finally, the physical process underpinning genetic inheritance – an idea proposed by Mendel almost a century prior – had been uncovered in all its complex glory.

Genes can exist in more than one form, or 'allele' – for example, the gene responsible for eye colour has different alleles for blue or brown eye colour.

James Watson (left) and Francis Crick (right) with a model of the double helix.

Who's Who – Rosalind Franklin

For decades, the name Rosalind Franklin went relatively unrecognised among the general public; but in recent years there has been more acknowledgement of the vital contributions made by 'the Dark Lady of DNA'.

Prodigiously intelligent, Franklin's family was prosperous, well connected and had a strong ethos of public service, and the young Franklin grew up in London with a profound appreciation of the value of knowledge and enquiry. In 1938, Franklin went up to Cambridge to study chemistry, and after graduation she made fundamental advances in research on the physics and chemistry of coal.

But Franklin's most legendary achievement was her work on DNA. In 1947, she moved to Paris to study a pioneering new technique known as X-ray crystallography, which was used to study the atomic structure of materials. In 1951, Franklin joined King's

Rosalind Franklin helped to crack the genetic code with her work on DNA.

College in London to work on the molecule underpinning life: DNA. At King's, Franklin's direct approach and fierce intelligence challenged her predominantly male colleagues, contributing to a difficult collaboration with Maurice Wilkins. In 1952, without Franklin's knowledge, Wilkins shared her key research finding – Photo 51, a diffraction image of DNA – with rival researchers James Watson and Francis Crick. In 1953, Watson and Crick announced that they had solved the structure of DNA. This was in no small part thanks to Franklin's work, but the pair failed to acknowledge her vital contributions either at the time or in their Nobel Prize acceptance speech some years later. Franklin died shortly after the discovery of DNA. Her untimely end was the result of ovarian cancer; she was just 37.

Rosalind Franklin is now rightly remembered as one of the greatest scientists of the 20th century. What's more, she is also recognized and celebrated as a powerful symbol for the unique hardships that women in science have faced throughout history. The censure, the erasure; Franklin was able to prosper despite the many gender-based obstacles she faced, and for that she remains a hero to many.

But there was still more to learn. In the late 1960s and early 1970s, Swiss and American microbiologists discovered an enzyme in bacteria that helped to protect the cell from attack. If a bacterium recognized unfamiliar DNA in its cells, it destroyed it by 'cutting' it into pieces. The 'restriction enzyme' was responsible for cutting invading DNA; it was sort of like a pair of molecular scissors. The discovery of these molecular scissors opened up an entirely new world for geneticists. For the first time, it became possible to cut and splice fragments of DNA from one place to another, and a few short years later, a team of American scientists did just that. In 1973, Stanley Cohen and Herbert Boyer carried out the first transplant of genes from one living organism into another; the era of genetic engineering had begun.

Throughout the following decades, genetic engineering was used to create a variety of new medicines, from synthetic insulin to vaccines; meanwhile, crops and livestock were developed to possess advantageous qualities such as pest resistance. In the early 21st century, genetic engineering entered a new age with the advent of CRISPR technology, which allowed for much more specific, straightforward and cost-effective gene editing.

These advances in gene editing tools have since supported the growth of a relatively new field of medicine known as gene therapy, which involves inserting foreign genes into a patient's cells and tissue in order to treat disease. The process usually involves healthy genes being used to replace abnormal, faulty genes, but it can also be used to deactivate mutated genes. Gene therapy was first successfully trialled in 1990, on a 4-year-old girl suffering from a genetic immune system deficiency, and researchers have since continued to develop the science. The pace of research into gene therapy is currently very rapid, and the field – while new and risky – offers much promise.

The same year, 1990, also saw the launch of a major international drive to sequence the entire human genome. DNA sequencing can help us to understand how our genes function, and opens up new avenues of research in biomedicine. The Human Genome Project – which effectively created a model blueprint for the genetic code of a human being – was completed in 2003, and the sequences of the entire genome are today publicly available for anyone to see. Since then, advances in genome sequencing have led to the decoding of the genome of Neanderthals, various animals, and even diseases such as lung cancer and malignant melanoma. The consequence of this enhanced understanding of organisms' genetic codes is that we're moving closer towards an era in which medicine can be tailored to an individual's genome.

While the concept of personalized medicine has existed in some form for many years, research really began to take off in the early decades of the 21st century. Cheaper and easier tools for synthesizing DNA have helped to support the drive to integrate genomics into patient care, from diagnostics through to therapy. Were personalized medicine to become widely viable, doctors could quickly identify genetic health risks in their patients, and could provide patients with the most effective, targeted treatments for their unique body. Personalised medicine is already making great strides in oncology, prenatal screening and pharmacology, but there's still much to learn about the intersections between healthcare and genetics, and there's a long way to go before genomics becomes a fully integrated clinical tool.

Genetics is still a relatively young field, but research is moving at breakneck speed. Of course, with great power comes great responsibility, and it's vital that we have rigorous oversight and monitoring of genetic research to ensure that advancements are delivered responsibly. Ultimately, we're at the very cusp of a revolution in genetics that could trigger the dawn of a new era in medicine and human health. There is no doubt that scientists will soon be able to unlock the transformative potential locked within our genes; the question is no longer 'if', but 'when'.

PART SIX

THE FUTURE OF SCIENCE

When we study the history of science, it seems almost fated that scientists will eventually arrive at the 'correct' conclusion – whether it's the model of the atom, the function of DNA, or the significance of the elements. With the benefit of hindsight, it's tempting to view science as a steady march of progress but, in reality, many of the accomplishments we take for granted took centuries of false starts, wrong turns and debate. In addition, of course, science is by no means complete – there are plenty of stories still being shaped, mysteries in the process of being unravelled, and answers waiting to be discovered. What might future historians of science say when looking back on the 21st century? What breakthroughs and seminal moments might we anticipate taking place over the course of the next century? Let's take a look at some of the most promising, challenging and potentially transformative areas of 21st century science – where do we expect to see advances, and what problems might we solve in the years to come? Of course, looking ahead is never the same as looking back, because, for us, that portion of history is still being written.

CHAPTER 17
LOOKING AHEAD

'Science is the key to our future.'

– BILL NYE (b. 1955), SCIENCE EXPLAINER,
PRESENTER AND MECHANICAL ENGINEER

ENERGY AND ENVIRONMENT

The future of science and the future of the planet are deeply interwoven, not least because we're currently in the midst of an environmental crisis. Scientists of myriad backgrounds – from geologists to computer scientists, chemists to epidemiologists – are currently investigating innovative technologies to meet environmental challenges. Foremost among these challenges is climate change.

At the time of writing, greenhouse gas levels are at an all-time high, and the UN Intergovernmental Panel on Climate Change (IPCC) recently warned that the effects of global warming could be irreversible by 2030. The consequences of this are potentially catastrophic – over 1 million species face extinction, and extreme weather conditions are predicted to displace up to 200 million people, creating huge numbers of climate refugees.

So should we anticipate any dramatic scientific breakthroughs waiting to save us from the climate crisis in the decades to come? The short answer is 'no'. Science and technology cannot solve climate change alone; preventing the disastrous consequences of global warming will take legislative action, geopolitical collaboration and widespread behavioural change. That said, innovation does have an important role to play in tackling the problem.

Technologies which enable us to reduce our reliance on fossil fuel are a key area of research, with advances in wind, solar, hydroelectric and nuclear energy sources all holding promise for the future. Many of these improvements will be iterative, with renewable technologies becoming cheaper and more efficient over time. But along the way, certain ingenious inventions are likely to stand out in the annals of history.

I. Research in Action

In the 21st century, research into more adaptive and efficient sources of renewable energy will be key. One such area of study is in thin-film solar cells, a potentially transformative technology that is likely to take off within the next few decades. You'll have seen thin films before: on your calculator, for example, a thin film is that small strip of material at the top that soaks up solar energy and helps to power the display. Thin films have been around since the 1980s, but in the early years of the 21st century, a new generation of thin films made from perovskite shot on to the scene. Regular solar panels are generally between 15 to 22 per cent effective, a statistic that refers to the percentage of power from the Sun that they convert into energy. In 2006, when early studies into perovskite

Science alone will not be able to solve climate change, but its dire consequences will eventually demand geopolitical collaboration, for example in industrial processes, and widespread behavioural change.

solar cells began, engineers reported around 3 per cent efficiency; by 2018, after years of refinements, the figure had risen to over 23 per cent. Unlike traditional solar materials, perovskite cells are soluble, meaning they can be made into a liquid and sprayed or painted on to a surface. In the future, we could see perovskite paints for the surfaces of houses and cars, and even clothing.

So why isn't this transformative solar material available to buy on the market? In addition to the usual challenges inherent in making a product scalable for widespread production, the big issue with perovskite cells is stability. The strength of the cells – their solubility – is also their weakness. Tests show that the material breaks down much more quickly than traditional solar panels, making them poorly suited to long-term energy supply. So the focus of research over the coming years will be to make perovskite cells more durable and better able to withstand the elements. But these challenges are not insurmountable, and it's likely that the nature of solar materials, along with other innovative sources of green energy, will continue to change dramatically throughout the 21st century, ushering in new approaches to renewable energy.

PIONEERING MEDICINE

Over the centuries, we've seen medicine become increasingly complex and in-depth, and that trend is likely to continue throughout the 21st century. In the last 100 years, we've seen human health improve the world over. In the UK, the average life expectancy in 1920 was a little over 57; a century later, that figure has risen to over 80, and in the years to come, it is expected to rise to an average of 90 in some countries. At the same time, vaccines, antibiotics and public health initiatives have dramatically altered the medical landscape. At the turn of the 20th century, respiratory and infectious diseases were by far the leading causes of death in the UK; at the turn of the 21st century, an individual's chances of dying from an infection have been slashed dramatically.

These medical advances were all hard won, and the product of highly sophisticated research into how organisms function at the smallest imaginable level. In the last 100 years, scientists have unpicked the submicroscopic defence mechanisms of bacteria, unravelled the mysteries of DNA, and revealed the invisible molecular workings of the human body – and we have these discoveries to thank for much of the good health we enjoy today. In the next 100 years, we can expect researchers to continue progressing our understanding of and control over processes such as health, disease and ageing by studying ever smaller and more complex biological units. Just like in the 20th century, these discoveries will certainly be aided by improvements in technology that will allow scientists to peer ever deeper into the structures and functions of life.

We've come a long way, but there's still a lot we don't understand about how the body works. In the 21st century, we can expect to see some of these great mysteries unpicked, spurring new innovations in healthcare.

II. Research in Action

They may seem like old news – after all, stem cells were first properly described all the way back in the 1960s – but stem cells have the potential to transform healthcare in the future. Their promise is that they could support regenerative medicine, reverse birth defects, and provide new treatments for diseases such as cancer, heart disease, diabetes and Parkinson's.

Stem cells are found in virtually all animals, including humans. They come in two different types: embryonic and adult. In adult organisms, stem cells help to repair and maintain the body. They renew tissue, dividing to create new skin, blood or bone cells, depending on the type of cell the body needs. Adult stem cells are specialized, but embryonic stem cells can turn into virtually any other kind of cell – they're not specialized in any way. The cells are formed in embryos 3–5 days old, and what type of cells they develop into is informed by complex signals.

Both types of stem cells have promising applications in medicine, but embryonic stem cells are the more versatile of the two because of their ability to freely transform into any type of cell. Scientists can create stem cell lines using embryos that have been discarded in IVF treatments – many embryos are grown for IVF, but only a handful are used, so they are often donated to science. Scientists have also started to discover ways of genetically altering adult stem cells to reprogram them into embryonic cells, thus increasing their medical potential – an achievement that earned the researchers involved a Nobel Prize in 2012. Although this research is still in its early stages, in recent experiments, animals with heart failure that were injected with these reprogrammed stem cells experienced improved heart function. In another recent study, reprogrammed stem cells were used to cure mice from a genetic mutation causing sickle cell anaemia.

Stem cells are likely to transform medicine over the course of the 21st century, and we're already seeing some signs of that potential being realized. Adult stem cells are being used in certain medical treatments, including leukaemia and heart disease, and the potential for reprogrammed stem cells is even greater. In the coming years, scientists will be working to learn more about the biochemistry behind stem cell development. By the end of the century, it's possible, if not likely, that stem cell technology will be a powerful and widely used clinical tool, transforming medical care.

NEXT-GENERATION TECH

In the 20th century, new technology transformed everyday life, leaving an irrevocable impact on modern society. The average person at the dawn of the 20th century would be astonished by the technologies we take for granted in the 21st century. From the Internet to telecommunications, portable gadgets to household goods, technology is deeply woven into the fabric of our lives in a way that no one could have predicted 100 years ago.

So what might we expect from the next 100 years when it comes to game-changing tech? It's likely that we'll continue to see trends towards greater automation. Processes that are currently led by human beings – such as controlling vehicles, manufacturing complex products, and even writing books – could be taken over by machines by the end of the 21st century. Society will need to adapt to these changes, reskilling workers or potentially adjusting the concept of work to allow for increased leisure time. Back in the 1930s, the British economist John Maynard Keynes predicted that technological improvements would lead to most people working only a 15-hour week. We haven't seen that transformation take place yet – if anything, people are working longer hours than ever before

– but the promise of technology is that it can make our lives easier, and potentially transform societies and economies at the same time. Perhaps Keynes' prediction will be realized by the end of the 21st century – only time will tell. What is for sure, though, is that the digital revolution is likely to continue for some years to come and, as we increase our mastery over materials, engineering and processes, it's likely that technology will become even more inextricably engrained within our everyday lives.

III. Research in Action

Artificial intelligence (AI) is one of the most exciting and potentially transformative technologies of the 21st century, yet its roots go back many years. Before AI was a reality, the British computer scientist Alan Turing established a principle to assess whether a machine was intelligent. Turing proposed that a human should be able to have a conversation with a machine without being able to tell that they were talking to a computer. While several versions of AI have come close, no computer has yet passed the Turing test.

In practice, AI has been an area of computer research since the 1950s, but we've really seen the field gather pace in recent years. In fact, AI is already integrated into our everyday lives, from Facebook face recognition to Netflix recommendations. But we've got a long way to go before we reap the transformative potential of intelligent machines.

Companies including Tesla and Google are currently trialling AI in self-driving cars. While there are a lot of issues that still need to be worked out, it's certainly not unthinkable to suggest that artificial intelligence could be standard in road vehicles by the end of the 21st century. And, of course, the potential for AI goes far beyond the automotive industry. Researchers are currently looking at ways of exploiting artificial intelligence for medical applications, particularly in the area of diagnosis. We're already starting to see signs of these developments; for instance, AI is currently being used in some medical settings to identify tuberculosis from radiography images, and this technology was recently found to perform better at diagnosing problematic skin lesions than human dermatologists.

The advance of AI poses interesting social, legal and even moral questions. It's important that, as we look to develop this technology, we do so responsibly. That said, the potential for AI is clearly huge, and the sense of excitement within the research community is palpable. By the end of the 21st century, we may well see AI applications in everything from the entertainment industry to fundamental scientific research.

EXPLORING THE UNIVERSE

Over the course of the 20th century, the range of human exploration increased dramatically. In the early years of the 1900s, it would have been hard to imagine that human beings would successfully depart the planet for space within just a few short decades. But a confluence of factors – political, scientific and profit-driven – have since triggered a new era in human exploration.

That said, space travel is currently far from commonplace. At the turn of the 21st century, fewer than 30 people had ever travelled beyond Earth's orbit. Even now, well over half a century since the USSR launched the first human into space, venturing out into the cosmos is still fraught with danger – and that's not to mention the astronomical expense of spaceflight.

Despite the very clear challenges inherent in space exploration, 21st-century technologies are already supporting new, more ambitious missions. Advances in robotics will certainly continue to be central to space exploration in the 21st century. The potential of space robotics is evident in the successes of the 'Curiosity' and 'Opportunity' rovers, which have helped scientists to explore the surface of Mars in unprecedented detail. The importance of these robotic astronauts is likely to become even more pronounced as we attempt to explore even more extreme environments and far-flung reaches of the cosmos.

One of the most hotly contested topics in space science at the moment is the concept of terraforming: that is, physically altering the living conditions on another planet to make it inhabitable for humans. While it's unlikely that we'll see a mass-migration of humans terraforming other areas of the solar system within the next 100 years, it's not impossible that we might achieve a small colony on Mars by the end of the 21st century. And of course, space tourism, for those who just fancy a passing visit, is almost certainly going to increase. The proliferation of private companies investigating space travel suggests that this could prove an increasingly popular – and lucrative – business in the years to come. So, while space will undoubtedly remain the final frontier in the year 2100, our solar system may be just a little less mysterious by then.

IV. Research in Action

While the early years of space exploration were focused squarely on the Moon, the modern space industry has set its sights firmly on Mars. Human beings have never set foot on the red planet; we've never even come close. In fact, since the first attempt back in the 1960s, around two out of every three attempted missions to Mars have failed. Why? Well, first off, Mars is very far away – about 140 million miles away, to be precise. That's around 200 times further than the Moon. It's also very hot, with high atmospheric pressure and frequent dust storms. We've never so much as managed to return samples from Mars to study on Earth – that's how difficult it is to explore the red planet.

That said, improvements in technology are constantly helping to push the boundaries of knowledge, and it's likely that the 21st century will bear witness to some major advances in Mars exploration. In 2018, NASA launched a pioneering new probe tasked with the mission of exploring beneath the internal surface of Mars. Known as 'InSight', the probe has already helped us to study seismic events on the red planet, known as 'Marsquakes'. Over the years to come, missions like this will be vital in helping us piece together the puzzle that is our closest planetary neighbour.

While there is a great deal of excitement about terraforming Mars for human habitation, we're still a long way from interplanetary colonization. Having said that, the knowledge we are currently gaining about the red planet is likely to inform future manned missions to Mars within the 21st century; and it's entirely possible that, in time, we will develop an extraterrestrial outpost on the planet. Are we likely to see those developments take place over the course of the 21st century? Ultimately, it's just too soon to tell.

CONCLUSION

ULTIMATELY, everything we can say about the future of science is little more than conjecture. No one really knows what the historians of the 22nd century might say about our little moment in scientific history. In the same way that thinkers at the dawn of the previous century couldn't possibly have predicted the wonders of the Internet, it's very difficult for us to imagine the ways in which science and technology will reshape everyday life in the years to come.

But what we do know is this: the journey towards those breakthrough discoveries and paradigm shifts will not be straightforward, because nothing in science ever is. There will be wrong turns, missteps, confusion and controversy. There will probably be periods where we take a step backwards; perhaps there will be moments of fundamental reckoning and scientific revolution – because that is how science works; that is how people work.

We curious beings – *homo sapiens*, the 'wise man' – are not simple creatures. The astonishing discoveries we have made are not the inevitable products of a triumphant march towards progress. In fact, much of the knowledge we now possess was once resisted, suppressed, ignored, or even lost. None of us, not even the supposed greatest minds of science, are entirely immune to politics, social pressures and emotional conflict. So ultimately, the story of science is not a tale of cold hard facts; it is, in fact, a very human tale – a messy, complicated, challenging story of a tribe bound together by shared curiosity, voyaging on an innate, instinctual quest to understand the nature of their being.

This is our story. Long may it continue.

FURTHER READING

Arney, Kat: *Herding Hemingway's Cats: Understanding How Our Genes Work*, Bloomsbury, London, 2017

Bryson, Bill: *A Short History of Nearly Everything*, Black Swan, London, 2004

Bynum, William: *A Little History of Science*, Yale University Press, London, 2013

Gribbin, John & Gribbin, Mary: *Science: A History in 100 Experiments*, William Collins, London, 2016

Hawking, Stephen: *A Brief History of Time: From The Big Bang to Black Holes*, Bantam Press, London, 2011

Henderson, Mark: *The Geek Manifesto: Why Science Matters*, Bantam Press, London, 2012

Ignotofsky, Rachel: *Women in Science: 50 Fearless Pioneers Who Changed the World*, Wren & Rook, 2017

Jha, Aloc: *How to Live Forever: and 34 Other Really Useful Uses of Science*, Quercus, London, 2012

Mun-Keat, Looi, Birch, Hayley & Stuart, Colin: *The Big Questions in Science*, Andre Deutsch, London, 2015

Mun-Keat, Looi & Stuart, Colin: *The Geek Guide to Life: Science's Solutions to Life's Little Problems*, Andre Deutsch, London, 2017

Rutherford, Adam: *The Book of Humans: The Story of How We Became Us*, W&N, London, 2018

Sagan, Carl: *Cosmos: The Story of Cosmic Evolution, Science and Civilisation*, Abacus, London, 1983

Tsjeng, Zing: *Forgotten Women: The Scientists*, Cassell, London, 2018

Yong, Ed: *I Contain Multitudes: The Microbes Within Us and a Grander View of Life*, Vintage, London, 2017

Zimmer, Carl: *Evolution: The Triumph of an Idea from Darwin to DNA*, Arrow, London, 2003

INDEX

PICTURE CREDITS

AKG Images: 62

Alamy: 19, 101, 114, 132, 174, 212, 215

Bridgeman Images: 18, 31, 38

Getty Images: 149, 191, 203, 240

Vittorio Luzzati: 230

Metropolitan Museum of Art, New York: 46, 82, 98

NASA: 112

National Museum of Archaeology, Athens: 210

Science & Society Picture Library: 25

Science Photo Library: 229

Shutterstock: 23, 37, 76, 141, 143, 144

Vassar College Library, Poughkeepsie, New York: 214

Wellcome Collection: 50, 88, 103, 106, 117, 130, 133, 166, 167, 172, 186, 187, 199, 201, 225, 227

Wikimedia Commons: 28, 35, 40, 59, 61, 64, 78, 87, 110, 123, 125, 127, 145, 153, 157, 158, 169, 183, 196, 217

David Woodroffe: 160, 162

Yale University, New Haven, Connecticut: 205